W9-BYX-629

THE ENDS OF THE WORLD

THE ENDS OF THE WORLD

Volcanic Apocalypses, Lethal Oceans, and Our Quest to Understand Earth's Past Mass Extinctions

PETER BRANNEN

ecco
An Imprint of HarperCollins*Publishers*

All insert photographs are courtesy of the author unless otherwise noted.

HarperCollins books may be purchased for educational, business, or sales promotional use. For information please e-mail the Special Markets Department at SPsales@harpercollins.com.

FIRST EDITION

Designed by Paula Russell Szafranski

Frontispiece © Horia Bogan/Shutterstock

Library of Congress Cataloging-in-Publication Data has been applied for.

ISBN 978-0-06-236480-7

17 18 19 20 21 LSC 10 9 8 7 6 5 4 3 2 1

To Mom

Something more than death has happened. . . . We are looking upon the uttermost finality which can be written, glimpsing the darkness which will not know another ray of light. We are in touch with the reality of extinction.

—Henry Beetle Hough

As often as I have seen beds of mud, sand, and shingle, accumulated to the thickness of many thousand feet, I have felt inclined to exclaim that causes, such as the present rivers and the present beaches, could never have ground down and produced such masses. But, on the other hand, when listening to the rattling noise of these torrents, and calling to mind that whole races of animals have passed away from the face of the earth, and that during this whole period, night and day, these stones have gone rattling onwards in their course, I have thought to myself, can any mountains, any continent, withstand such waste?

—Charles Darwin

CONTENTS

THE ENDS OF THE WORLD

INTRODUCTION

It is the dawn of a new geological age. A teeming swarm of *Homo sapiens* gathers on the banks of an estuary at the edge of the North American continent. The glaciers have retreated; the seas have risen more than 400 feet since the last ice age; and the gleaming new steel-and-glass hives of Manhattan rise up from the marshes. Looming over the confident city, just across the Hudson River, is the sheer cliff face of the Palisades. The gigantic columns of basalt sit in unimpressed, stony silence, as they have for 200 million years. These cliffs, covered in highway weeds and graffiti, are monuments to an ancient apocalypse. They're made of magma that once fed burbling fountains of lava at the surface—lava that once smothered the planet from Nova Scotia to Brazil. The eruptions flooded the atmosphere with carbon dioxide at the end of the Triassic period, roasting the planet and acidifying the oceans for thousands of years. Brief blasts of volcanic smog punctuated this super-greenhouse with cold. The runaway volcanism covered more than 4 million square miles of the planet and killed off more than three-quarters of animal life on earth in a geological instant.

I struggled to keep up with Columbia University paleontologist Paul Olsen as he bounded up the scraggly path leading from the banks of the Hudson to the base of the Palisades. In front of us, smothered under this enormous wall of now-solid magma, were the remains of a quarter-billion-year-old lake bottom, complete with exquisitely preserved fish and reptile fossils. Behind us, faintly droning, was the skyline of New York City.

I asked Olsen whether the city across the river would be preserved for future geologists to discover, like this peaceful Triassic diorama at the bottom of the rocks. He turned to consider the scenery.

"You might have a layer of stuff," he said dismissively, "but it's not a sedimentary basin, so eventually it would erode away to nothing. You'd have bits that would make it out into the ocean and would be buried and might show up—some bottle caps, maybe. There would be some pretty heavy-duty isotopic signals. But the subway system wouldn't fossilize or anything. It all would erode away fairly quickly."

It is from this disorienting perspective that geologists operate: to them, millions of years run together, seas divide continents, then drain away, and great mountain ranges erode to sand in moments. It's an outlook that's necessary to cultivate if one wants to get a handle on the staggering depths of geological time, which recedes behind us hundreds of millions of years and stretches out before us to infinity. If Olsen's attitude seems dispassionate in the extreme, it's a symptom of a lifetime's immersion in Earth's history, which is both vast beyond comprehension and, in some exceedingly rare moments, tragic beyond words.

Animal life has been all but destroyed in sudden, planetwide exterminations five times in Earth's history. These are the so-called Big Five mass extinctions, commonly defined as any event in which more than half of the earth's species go extinct in fewer than a million years or so. We now know that many of these mass extinctions seem to have happened much more quickly. Thanks to fine-scale geochronology, we know that some of the most extreme die-offs in earth history lasted only a few thousand years, at the very most, and may have been much quicker. A more qualitative way to describe something like this is Armageddon.

The most famous member of this gloomy fraternity is the End-Cretaceous mass extinction, which notably took out the (nonbird) dinosaurs 66 million years ago. But the End-Cretaceous is only the most *recent* mass extinction in the history of life. The volcanic doomsday whose stony embers I saw exposed in the cliffs next to Manhattan—a disaster that brought down an alternate universe of distant crocodile relatives and global coral reef systems—struck 135 million years before the death of the dinosaurs. This disaster and the three other major mass extinctions that preceded it are invisible, for the most part, in the public imagination, long overshadowed by the downfall of *T. rex*. This isn't entirely without reason. For one thing, dinosaurs are the most charismatic characters in the fossil record, celebrities of earth history that paleontologists who work on earlier, more neglected periods scoff at as preening oversized monsters. As such, dinosaurs hog most of the popular press spared for paleontology. In addition, the dinosaurs were wiped out in spectacular fashion, with their final moments punctuated by the impact of a 6-mile-long asteroid in Mexico.

But if it *was* a space rock that did in the dinosaurs, it seems to have been a unique disaster. Some astronomers outside the field push the idea that periodic asteroid strikes caused each of the planet's other four mass extinctions, but this hypothesis has virtually no support in the fossil record. In the past three decades, geologists have scoured the fossil record looking for evidence of devastating asteroid impacts at those mass extinctions, and have come up empty. The most dependable and frequent administrators of global catastrophe, it turns out, are dramatic changes to the climate and the ocean, driven by the forces of geology itself. The three biggest mass extinctions in the past 300 million years are all associated with giant floods of lava on a continental scale—the sorts of eruptions that beggar the imagination. Life on earth is resilient, but not infinitely so: the same volcanoes that are capable of turning whole continents inside out can also produce climatic and oceanic chaos worthy of the apocalypse. In these rare eruptive cataclysms the atmosphere becomes supercharged with volcanic carbon dioxide, and during the worst mass extinction of all time, the planet was rendered a hellish, rotting sepulcher, with hot, acidifying oceans starved of oxygen.

But in other earlier mass extinctions, it might have been neither volcanoes nor asteroids at fault. Instead, some geologists say that plate tectonics, and perhaps even biology itself, conspired to suck up CO_2 and poison the oceans. While continental-scale volcanism sends CO_2 soaring, in these earlier, somewhat more mysterious extinctions, carbon dioxide might have instead plummeted, imprisoning the earth in an icy crypt. Rather than spectacular collisions with other heavenly bodies, it has been these internal shocks to the earth system that have most frequently knocked the planet off course. Much of the planet's misfortune, it seems, is homegrown.

Luckily, these uber-catastrophes are comfortingly rare, having struck only five times in the more than half a billion years since complex life emerged (occurring, roughly, 445, 374, 252, 201, and 66 million years ago). But it's a history that has frightening echoes in our own world—which is undergoing changes not seen for tens of millions, or even hundreds of millions, of years. "[It's] pretty clear that times of high carbon dioxide—and especially times when carbon dioxide levels rapidly rose—coincided with the mass extinctions," writes University of Washington paleontologist and End-Permian mass extinction expert Peter Ward. "Here is the driver of extinction."

As civilization is busy demonstrating, supervolcanoes aren't the only way to get lots of carbon buried in the rocks out into the atmosphere in a hurry. Today humanity busies itself by digging up hundreds of millions of years of carbon buried by ancient life and ignites it all at once at the surface, in pistons and power plants—the vast, diffuse metabolism of modern civilization. If we see this task to completion and burn it all—supercharging the atmosphere with carbon like an artificial supervolcano—it will indeed get very hot, as it has before. The hottest heat waves experienced today will become the average, while future heat waves will push many parts of the world into uncharted territory, taking on a new menace that will surpass the hard limits of human physiology.

If this comes to pass, the planet will return to a condition that, though utterly alien to us, has made many appearances in the fossil record. But warm times aren't necessarily a bad thing. The dinosaur-haunted Cretaceous was significantly richer in atmospheric CO_2, and that period was consequently much warmer than today. But when climate change or ocean chemistry changes have been sudden, the result has been devastating for life. In the worst of times, the earth has been all but ruined by these climate paroxysms as lethally hot continental interiors, acidifying, anoxic oceans, and mass death swept over the planet.

This is the revelation of geology in recent years that presents the most worrying prospect for modern society. The five worst episodes in earth history have all been associated with violent changes to the planet's carbon cycle. Over time, this fundamental element moves back and forth between the reservoirs of biology and geology: volcanic carbon dioxide in the air is captured by carbon-based life in the sea, which dies and becomes carbonate limestone on the seafloor. When that limestone is thrust down into the earth, it's cooked and the carbon dioxide is spit out by volcanoes into the air once more. And on and on. This is why it's a cycle. But events like sudden, extraordinarily *huge* injections of carbon dioxide into the atmosphere and oceans can short-circuit this chemistry of life. This prospect is one reason why past mass extinctions have become such a vogue topic of late in the research community. Most of the scientists I spoke with over the course of reporting this book were interested in the planet's history of near-death experiences, not just to answer an academic question, but also to learn, by studying the past, how the planet responds to exactly the sorts of shocks we're currently inflicting on it.

This ongoing conversation in the research community is strikingly at odds with the one taking place in the broader culture. Today much of the discussion about carbon dioxide's role in driving climate change makes it seem as though the link exists only in theory, or in computer models. But our current experiment—quickly injecting huge amounts of carbon dioxide into the atmosphere—has in fact been run many times before in the geological past, and it never ends well. In addition to the unanimous and terrifying projections of climate models, we also have a case history of carbon dioxide–driven climate change in the planet's geologic past that we would be well advised to consult. These events can be instructive, even diagnostic, for our modern crises, like the patient who presents to his doctor with chest pains after a history of heart attacks.

But there's a risk of stretching the analogy too far: Earth has been many different planets over its lifetime, and though in some salient and worrying ways our modern planet and its future prospects echo some of the most frightening chapters in its history, in many other ways our modern biocrises represent a one-off—a unique disruption in the history of life. And thankfully, we still have time. Though we've proven to be a destructive species, we have not produced anything even *close* to the levels of wanton destruction and carnage seen in previous planetary cataclysms. These are absolute worst-case scenarios. The epitaph for humanity does not yet have to include the tragic indictment of having engineered the sixth major mass extinction in earth history. In a world sometimes short on it, this is good news.

Like many kids, I came to the topic of mass extinctions early. As the son of a children's librarian, I grew up in a house that was often brimming with cardboard boxes of books—the surplus of the most recent book fair. Perhaps to my mom's frustration, I would pass over copies of *Where the Red Fern Grows* and *The Giver* and go straight for the pop-ups. Tyrannosaurs and cycads leapt from the page as I obsessed over the strange Latinate names and the even stranger creatures they described. Here an artist had decided to spangle a bizarre-looking animal called *parasaurolophus* in neon, while another illustrator had oviraptors draped in zebra stripes. It was irresistible: a world of sci-fi monsters that had actually existed. But Disney's *Fantasia* illuminated for me as a child an even stranger fact about this world: that it had all occurred in the past—to the music of Stravinsky's orchestra, the dinosaurs lurched to their deaths over a cauterized landscape, and the world ended in tragedy. It was no more. Later obsessions—like the movie and book versions of *Jurassic Park*—only reinforced for me the melancholy of living in a world that had lost its dragons.

In the past few decades, geologists have started filling in the rough sketches of the Big Five mass extinctions with gruesome detail, but the story has largely eluded the public imagination. Our conception of history tends to stretch back only a few thousand years at most, and typically only a few hundred. This is a scandalously shortsighted appreciation of what came before—like reading only the last sentence of a book and claiming to understand what's in the rest of the library. That the planet has nearly died five times over the past 500 million years is a remarkable fact, and as we, as a civilization, push the chemistry and temperature of the climate-ocean system into territory not seen for tens of millions of years, we should be curious about where the hard limits are.

Just how bad could it get? The history of mass extinctions provides the answer to this question. Visiting Earth's turbulent and unfamiliar past provides a possible window into our future.

Forgotten worlds spill from the sides of highways, from beach cliffs, and from the edges of baseball fields, hiding in plain sight. This was perhaps the central revelation to me as I began to accompany paleontologists in the field to learn more about the five major mass extinctions. I didn't have to talk my way onto expeditions to the Arctic or the Gobi Desert to find the strange stratigraphy of long-past worlds. We live on a palimpsest of earth history. The lesson of geology is that we inherit this world—this "antique planet with a brand new civilization," as Carl Sagan put it—from countless vanished ages. To see the world through the lens of geology is to see the world for the first time.

In North America, fossils are found not only in the mythic Southwest and in exposed Arctic mountainsides but hidden under Walmart parking lots, in quarries, and in road cuts on the interstate. Underneath Cincinnati is an endless fossil bas-relief of tropical sea life in the early oceans of the Ordovician period, which ended half a billion years ago in the second worst extinction in Earth's history. There are plesiosaurs in riverbanks in downtown Austin, saber-toothed cats in Los Angeles, and killer crocs from the Triassic under Dulles Airport outside of Washington, DC. In Cleveland's riverbanks are the armor-plated remains of a guillotine-mouthed, titanic fish from the Devonian period, 360 million years old.

The wreckage from the Big Five mass extinctions lies on remote, verdant islands in the Canadian Maritimes, on icier patches in Antarctica and Greenland, under Mayan temples in Mexico, strewn across the desolation of South Africa's Karoo Desert, and on the edges of farmland in China. But this legacy of disaster is also visible next to skyscrapers in New York City and in the shales of the Midwest (so profitable for frackers and environmental fundraisers alike) that were forged in the chaos of the Late Devonian mass extinctions. Rising out of the deserts of West Texas are the Guadalupe Mountains, a haunted monument built almost entirely from ancient sea animals in the full bloom of life before the single worst chapter in the planet's history: a period of crises capped by a carbon dioxide–driven global warming catastrophe that killed off 90 percent of life on Earth.

Life on earth constitutes a remarkably thin glaze of interesting chemistry on an otherwise unremarkable, cooling ball of stone, hovering like a sand grain in an endless ocean of empty space.

This sheet of life that coats the planet—a feature of our world that has been almost miraculously durable over Earth's history—is perhaps unique in the galaxy. But viewed through the lens of mass extinctions, it's also remarkably fragile: when crises push the planet outside a narrow set of surface conditions, it has been nearly sterilized. Much has been made of the search beyond our planet for spectacular external threats like asteroids, but we should be equally vigilant about the subtler threats from within. As the roster of lifeless planets in our solar system attests, the agreeable chemistry and conditions on the surface of the earth are incredibly unusual. And as the history of mass extinctions demonstrates, they're not a given.

In researching these ancient disasters, I expected to find a story as neat and tidy as the one about the asteroid that killed the dinosaurs. What I found instead was a frontier of discovery with much left to be unearthed, and a story still largely obscured by the fog of deep time. In my travels I became acquainted with whole worlds—still called "Earth"—that I had scarcely known existed, brought low by a suite of world-ending forces far subtler, but just as ominous, as asteroids.

This book is a woefully incomplete testament to the ingenuity of those who have labored to piece this fractured—and still unfinished—puzzle together, as well as a survey of the unfamiliar geography of deep time that surrounds us. It's also an exploration of the turbulent centuries to come and the long-term prospects for life on this strangely hospitable but vulnerable planet that hurtles through a perilous universe.

After hiking the Palisades, Olsen and I hit up one of the dozens of Vietnamese pho restaurants in the nearby Fort Lee neighborhood, where a snarl of highways branch out of the George Washington

Bridge. Contemplating the history of the region and the ancient hellscape created by the rocks underneath us, I found it difficult not to wonder about the future. Currently the carbon dioxide concentration in the atmosphere hovers at around 400 parts per million (ppm)—probably the highest it's been since the middle of the Pliocene epoch 3 million years ago. What will life be like on the planet at 1,000 parts per million, which some climate scientists and policymakers project for the coming decades if we continue to take a business-as-usual approach to emissions?

"The last time anything like that occurred, we had no polar ice at all and sea levels were hundreds of feet higher," Olsen said, noting that crocodiles and lemur-relatives inhabited the tropical *northern* shores of Canada. "Ocean temperatures in the tropics were possibly 40 degrees Celsius on average, which would be completely alien to us now.

"The interior of continents," he continued, "endured persistently lethal conditions."

I put the question a little more bluntly, asking him whether we might be at the beginning of another mass extinction.

"Yeah," he said, resting his chopsticks for a moment. "Yeah. Although the one that would be obvious in the fossil record happened over a 50,000-year interval from the time that humans spread out of Africa and wiped out all the megafauna. That's the one that will show up like gangbusters in the fossil record. Someday they might say that the industrial spread of humans was just the coup de grâce."

1

BEGINNINGS

We have a sense that the beginning of animal life on our planet was like a dawning spring. But the reality is that the age of animals was like a baby born to almost impossibly old parents.

—Peter Ward

I'm from Boston. Conveniently, this means it's only a short commuter ferry ride across the harbor to see what might be some of the earliest fossils of large, complex life in the history of the planet. Below a marina ringed with condos and the trappings of strip mall modernity is a beach studded with the rusty spikes of some bygone wharf. At the far end of the neglected beach, low tide reveals slabs of ancient seafloor draped in seaweed, sloping into the sea. The rocks, from the bottom of the ocean off the coast of a supercontinent near the South Pole, poke out not far from the Bed Bath & Beyond parking lot. They are more than half a billion years old. No plaque or marker indicates that there's anything particularly interesting about them, but brushing back the wrack reveals concentric ovals, no larger than a quarter, that pock the surface of the stone. The unassuming rings in the rock might mark the imprint where a fern-shaped creature anchored itself to the slimy silt at the bottom of the ocean at the dawn of complex life.

This is where the story starts. On a planet that shares our name, but that's about it.

It's impossible to grasp how long ago these creatures led their strange lives on the Antarctic Boston seafloor. It's doubly impossible to grasp how old the planet is—or how insignificant the role humanity has played on it. With his paean to "the Pale Blue Dot," Carl Sagan helped illustrate how utterly marooned we are in our tiny, far-flung corner of space. But we are similarly marooned in time, between incomprehensible eternities. Luckily, geologists have come up with some mental tricks to help us grok our place among the eons. One of them involves a footstep analogy* that goes something like the following: imagine each step you take represents 100 years of history. The simple conceit has stupefying implications.

Let's begin our walk; we'll start in the present and head back. As you lift up your heel there's no Internet, one-third of the earth's coral reefs reappear, atomic bombs violently reassemble, two world wars are fought (in reverse), the electric glow on the night side of the planet is extinguished, and—when your foot lands—the Ottoman Empire exists. One step. After twenty steps, you stroll by Jesus. A few paces later the other great religions begin to wink out of existence: first Buddhism, then Zoroastrianism, then Judaism, then Hinduism. With each footfall, the cultural milestones get more staggering. The first legal systems and writing disappear, and then, tragically, so does beer. After only a few dozen steps—before you can even reach the end of the block—all of recorded history peters out, all of human civilization is behind you, and woolly mammoths exist. That was easy. You stretch your

* Courtesy of the Carnegie Institution's Robert Hazen.

legs and prepare for what couldn't be much longer of a walk. Perhaps it's a short stroll to the dinosaurs, and a little farther still to the trilobites. No doubt you'll be at the formation of the earth by sundown. Not so.

In fact, you would have to keep walking for *20 miles a day, every day, for four years* to cover the rest of the planet's history.* Clearly the story of planet Earth is not the story of *Homo sapiens*. Almost all of that walk would be through a forbidding landscape with no complex life on it whatsoever. Not in the deep sea, not atop the mountains, not in the tropics, nor on the endless barren granite interiors of the continents. Save for the wind and the waves, ours was a silent planet for the most part during this nearly eternal preamble to animal life. Those first creatures, stamped in the rocks of Boston Harbor and elsewhere, came after 4 *billion* years on Earth without anything on the entire face of the planet more exciting than pond scum. In fact, the years between 1.85 billion and 850 million years ago were so uneventful that even geologists have taken to referring to them as the "boring billion." When a geologist calls something boring, reel in horror.

As we search for life around other planets this is something to keep in mind: even the earth was a desolate wasteland for 90 percent of its history. In fact, one of the only signs of life in the rock record for billions of years is the presence of uninspiring mounds of fossilized microbial slime. Then, around 635 million years ago, a tiny whisper of complex life: rocks found in Oman bear 24-isopropylcholestane, a mouthful of a chemical that today is produced only by certain sponges. Sponges got busy filtering the sea and burying carbon and thus may have ventilated the oceans, making more complex life possible. As the Smithsonian's Doug Erwin writes, "Humanity owes a special debt to sponges." Something to

* To reach the big bang you'd have to keep trudging at the same pace for almost another decade.

keep in mind the next time you're using one to wipe bacon grease off a pan.*

Then, around 579 million years ago, during the Ediacaran period, after a spell of near-sterilizing global ice ages (aptly called Snowball Earth),† the champagne bottle of life was uncorked and large, complex creatures finally, and rather suddenly, appear as fossils on the ancient ocean floor.

Although this is still recent history in the 4,500-million-year life span of the planet, it's still unspeakably old—more than 200 million years before the supercontinent Pangaea assembled, and more than *500 million years* before *T. rex*. And at 579 million years ago, it's about 579 million years before modern humans, whose years on this planet are measured in hundreds of thousands rather than millions. Even for geologists, these abysses of deep time far surpass all understanding.

The first simple creatures that suddenly appear in the fossil record were probably not animals at all. And their reign would be a short one. In fact, they might have endured the first mass extinction ever, leaving only their cryptic shapes in the rocks, their lives discernible only through the poetry of paleontologists.

Across the windswept "hyperoceanic barrens" of southeast Newfoundland, and not far from the lonely telegraph station that picked up the last distress signals of the HMS *Titanic*, is still more fossil graffiti left on old ocean rocks by these pseudo-creatures—hieroglyphic echoes of life in the perpetual midnight of the ancient deep. Some of the Newfoundland fossils recall fern fronds, feather dusters, and slender cones, while others appear as large, Seuss-like segmented slugs or bloated centipedes. They seem to have

* Fortunately, this sort of ancestral disrespect is rare: most kitchen sponges are synthetic.

† The planet was probably rescued from Snowball Earth by carbon dioxide from volcanoes, which warmed up the planet.

invented a way of life—a mostly immobile one—unlike anything alive today: sluggishly sucking up organic gunk in the disgusting seas of the primordial earth across their membranes. But this way of life was a failed attempt at life on earth. By the next age all these creatures would be gone.

Around 540 million years ago, the Ediacaran world was destroyed—dramatically swept aside in the most important moment in the history of evolution: the Cambrian Explosion. When this spectacular supernova of biology detonated, the world of animal life—creatures that move around and eat other organisms for a living—was truly born. Though there are fossil whispers of an emergent animal lineage in the staid age that came before, the turbid seas had been dominated until then by the almost inert, fractal pseudo-creatures of the Ediacaran period. That all changed at the dawn of the Cambrian. Animals rapidly diversified and overthrew this weird life with a menagerie of even weirder life. Though it hasn't been inducted into the ranks of the canonical Big Five mass extinctions, the Cambrian Explosion, counterintuitively, might have also marked the first such mass death in the history of complex life.

If the forgotten creatures of the Ediacaran period in Newfoundland and elsewhere look like graffiti left by aliens, then the flamboyant animals of the Cambrian Explosion that replaced them look like the aliens themselves. The seas were suddenly stocked with creatures that would be difficult to invent during the most frenzied acid trip—indeed, one Cambrian animal is even named *Hallucigenia*. Another, *Opabinia,* with five eyes and a bizarre armlike appendage where one would expect a mouth, drew peals of laughter when it was first described at a scientific meeting. Still others, like the iconically weird *Anomalocaris*—looking something like an undulating, satanic lobster—invite us to squint when imagining its place on our common tree of life. Their unrecognizable forms, now entombed in museum displays and tantalizingly rendered in artists' depictions, stand as reminders that, though it was still technically "Earth," this planet has been many altogether different worlds over its lifetime.

Some of these animal experiments were just that—experiments. And some experiments fail, never to be reproduced again. Others were more successful: on the bizarre roster of creatures from the Cambrian Explosion is an ancestor of ours, perhaps the unimpressive 2-inch, lancelet-like *Metaspriggina*.

The widespread appearance of animals starting in the Cambrian period is so startlingly abrupt in the fossil record that its seeming spontaneity worried Darwin. The more than a century of investigation since has shown that the explosion wasn't quite so instantaneous, but from a geological perspective it was still shockingly swift. The causes of the explosion are still hotly debated. They range from an increase in oxygen in the oceans (possibly a product of life itself), which would have underwritten the more energetic lifestyles of animals, to more speculative causes, like the invention of vision, which would have suddenly illuminated the zero-sum playing field for predators and prey, lighting the fuse of a predatory arms race. But lost amid the hubbub of the Cambrian Explosion is the sad story of the brief world that came before and whose mysterious, forgotten forms vanished forever. When animal life exploded, those strange fleshy fronds at the bottom of the ocean and bloated sluglike creatures would never be seen again.

"It was a mass extinction that was ultimately caused by the evolution of new behaviors," said Vanderbilt paleontologist and Ediacaran expert Simon Darroch. I caught up with Darroch at a geology conference in Baltimore. A fresh-faced and affable scientist who speaks the Queen's English, Darroch sticks out among the crowd of goateed, mildly autistic, middle-aged, midwestern American males that haunt stateside geology conferences.

The disappearance of the strange world that preceded the Cambrian Explosion—a zen garden world of unfamiliar fractal creatures rising from the seafloor and strange quilted blobs hugging the microbial mats—has long been a mystery to paleontologists. But in 2015, Darroch and his colleagues declared the cold case to be a mass extinction.

"We think of mass extinctions as requiring an abiotic driver: an asteroid impact or a period of volcanism. But here there's strong evidence that biological organisms that changed their environment drove the extinction of vast swaths of complex, eukaryotic life. I think it's a powerful analogy for what we're doing today."

One new behavior in particular seems to have been responsible for much of the disruption: burrowing. The strange geometric creatures in Newfoundland and elsewhere depended on revolting, organic-rich murky seas, along with seafloors paved with undisturbed microbial muck, to survive. But when the Cambrian Explosion went off and animals inherited the earth, they began churning up the seafloor. For the strange quilted blobs of the former Ediacaran period that sat on the bottom and absorbed the nutrition from the placid sheets of slime, this was catastrophic. In fact, burrows in the rocks officially define the start of the Cambrian period for geologists. They might have been left there by so-called penis worms (no kidding), which churned through the primeval seafloor and ruined the Ediacaran habitat. For geologists, the burrows mark a qualitative change in the strata, separating it from the billions of years of unburrowed rocks that come before. And it's a change perhaps unmatched in the rock record for the next half-billion years, until humans began leaving miles-deep holes in the rocks, in search of minerals and fossil fuels.

The animal arrivistes of the Cambrian Explosion also began filtering the seas and delivering ever more organic carbon that had been suspended in the water column to the seafloor. In other words, they started pooping. As a result, the strange fractal fronds of the preceding Ediacaran period were suddenly left suspended in a frightfully clear sea with nothing to eat.

The flip side of this new Cambrian animal menagerie taking all this carbon gunk out of the water and burying it in the seafloor might have been an even greater boost of oxygen in the ocean. This boost might have further fueled the arms race of innovation then escalating in the seas, leaving the poor sluggish pseudo-creatures behind. By ventilating the oceans, animal life was making the planet ever more habitable for more animal life and prodding ever crazier experiments in biology. What hope did a quilted blob or a motionless fractal frond have in a world that was weaponizing with tentacles and exoskeletons and claws?

A sentiment exists—particularly among nonscientists—that the idea of humans seriously disrupting the planet on a geological scale is mere anthropocentric hubris. But this sentiment misunderstands the history of life. In the geological past, seemingly small innovations have reorganized the planet's chemistry, hurling it into drastic phase changes. Surely humans might be as significant as the filter-feeding animals of the Cambrian Explosion.

"It's not mind-blowing stuff, but I think it's hard for people to accept because we don't see ourselves as that important in the grand scheme of things," Darroch said. "But here's an example where, 500 million years ago, something very similar happened. There's plenty of talk today about comparing rates of extinction in the mass extinctions of the past with the rate at which we're driving species extinct today, and it's all through the evolution of new behaviors and ecosystem engineering."

Like the Cambrian burrowers that reshaped the microbial mat world to their own ends, humans have converted half the planet's land surface to farmland. We're even beginning to change the chemistry of the ocean, acidifying it with carbon dioxide and turning whole swaths of the continental shelves anoxic with the deluge of nitrogen and phosphorus fertilizers pouring out of our agricultural heartlands. And the dizzying arsenal of our modern technology is a leap in innovation perhaps matched in the entire history of life only by the eruption of biological invention at the Cambrian Explosion. At the very least, it's not a stretch to think we might be as important as penis worms.

"So I just think, here's an example from the past when an ecological crisis happened because of ecosystem engineering," said Darroch. "And we shouldn't be too surprised or too staggered or too blown away by the fact that maybe it's happening again. Biological organisms are an incredibly powerful geological force."

The Cambrian Explosion—though it might have been devastating for the strange Ediacaran creatures that came before—was an unambiguously good thing for life on earth. It marked the official beginning of animals' stewardship over a planet long immersed in "the boring billions." Perhaps today the new technological world we have built for ourselves marks the beginning of a similarly epochal transition, and a new eon awaits that will appear as alien to us in 10 million years as the dizzying animal world of the Cambrian would have appeared to the pitiable creatures that came before. Or perhaps our shocks will prove less auspicious with humanity instead, leaving behind a ruined world—our legacy consisting only of a long environmental convalescence from civilizational excess.

As for the Cambrian period, *its* legacy was the tapestry of all animal life, which had unspooled from some forgotten ancestor. The planet was now an active one. Life crept and swam and spied on itself with eyes and chemoreceptors. Creatures killed one another and ate one another and hid in terror. Though we wouldn't recognize it whatsoever, this was now our world—red in tooth and claw. After a 4-billion-year prologue that started in fire and ended in Snowball Earth, the pageant of animal life had begun, and the next half-billion years would be the most interesting by far.

The Cambrian Explosion might get all the credit for launching animal life on earth, but the ocean of the Cambrian period remained impoverished for millions of years as pulses of anoxic water intruded into the shallows, wiping out species after species in wave after wave of extinctions. This strange retardation of life that followed the Cambrian Explosion has been ominously referred to as the "Cambrian Dead Interval." But the dark ages ended when the ensuing Ordovician period began. Almost until it ended, the next age would oversee an unprecedented roll of evolutionary good times.

The Ordovician period would be a riotous time for life on earth, with an incredible boom, unlike any in earth history, followed by an even more incredible bust. The age of mass extinctions had begun.

2

THE END-ORDOVICIAN MASS EXTINCTION

445 Million Years Ago

Snow had fallen, snow on snow,
Snow on snow,
In the bleak midwinter
Long ago.
—*Christina Rossetti, 1904*

It was Friday night at the University of Cincinnati, and the football stadium was rocking. A pack of undergrads boozily wandered through the darkened campus, led toward the beacon of Jumbotrons and floodlights spilling over the stadium walls. They ambled past the physics building in the shadows, unaware that, inside it, a somewhat less rowdy Friday-night ritual was also in progress. At the end of a dimly lit hallway, the lights were on, which could mean only one thing: the monthly meeting of the Dry Dredgers.

Though they might sound like some kind of off-brand Freemasons, the Dry Dredgers—one of the most respected amateur fossil-collecting groups in the country—are open to all comers. The only membership requirement is an obsession with deep time. They have been meticulously "dredging" the Cincinnati area for ancient sea life since 1942 in weekend fossil-hunting jaunts, with countless citations in paleontology papers to show for it. Because their home base is in southwest Ohio, the group sits atop bedrock made of an old ocean seafloor, and they specialize in fossils from the Ordovician period, an alien world that lasted from 488 million to 443 million years ago, and that ended in catastrophe.

The mild Ordovician world was suddenly destroyed at the end by a surprising ice age—and then punished again by a tide of noxious seas. The resulting mass extinction inflicted by these wrenching climate swings would be the second worst in the history of life.

Seeing as my enthusiasm for paleontology began where most people's did—namely, the time when the big, scaly things started to lumber around—I knew hardly anything about this *much* older planet, one that was still almost completely barren of life on land and would remain so for almost 100 million years. But ours has always been an ocean planet foremost, and there was no shortage of action below the waves in the Ordovician. So I came to the seas of Cincinnati for an introduction.

"Boy, everyone's doing show-and-tell tonight," said Dry Dredgers president Jack Kallmeyer before the meeting was called to order. His members milled about, showing off their spoils from the field and peering into one another's shoeboxes, filled with petrified creatures salvaged along roadsides or in old quarries since last month's meeting. The hard-core hobbyists—in town from all over the Midwest—traded war stories of collecting since their last get-together. They commiserated about lost fossil sites, shuttered by phosphate mining companies or plowed over by suburban subdivisions.

It's a common lament among rockhounds. Few real estate developers know or care about the fossils standing in the way of their next cul-de-sac, and most Americans have no idea that the thin veneer of civilization, with its strip malls, pavements, and dutifully irrigated patches of sod, mostly sits above a fathomless underworld of fossils. This reality might be more difficult to avoid in the Cincinnati area, which is propped up on top of a gigantic hash of ancient tropical sea life that literally spills out of the sides of the roads. The area, including neighboring northern Kentucky and southeastern Indiana, has been called one of "the most fossil-rich regions in North America, if not the entire world," and has been a magnet for paleontologists for almost 200 years. The city is so rich in fossils that it even has a chunk of earth history named after it: the Cincinnatian.*

After show-and-tell, the members took their seats. It was a decidedly older crowd, many of whom seemed to have more than an academic interest in the details of fossilization. The lecture that night came courtesy of an Illinois high school science teacher and fellow Paleozoic fossil nut who held court on an obscure line of stalked filter-feeders that arose during the Ordovician.

"When you talk about blastoids, you gotta talk about *pentremites,*" he said. I looked around as the group nodded along, endorsing the collaboration of these unfamiliar nouns. "But, by far, my favorite are the *diploblastus.*"

To audible "oohs" from the audience, he pulled up a slide on the overhead projector of an ambiguous fossilized bauble.

---

* The Cincinnatian series is the latest part of the Ordovician series in North America.

"Here's *Tricoelocrinus woodmani,*" he said. "The Rolls-Royce of blastoids."*

The man sitting in front of me was wearing a T-shirt solemnly inscribed not with an inspirational quote, but with the text of the mayoral proclamation that Cincinnati's official fossil would be *Isorophus cincinnatiensis,* an extremely distant relative of starfish that lived over 400 million years ago. These guys were amateurs in name only.

At the end of the night, Kallmeyer handed out the itinerary for the weekend: we would be joining the Kentucky Paleontological Society the next morning and then set out to sea.

Bleary-eyed, I joined a convoy of cars just outside the city the next day. The first stop—tucked away at the end of a service road—was an exposed hillside of layered gray rock, of the same sort that rises from the sides of the local highways. After we scampered up the bluffs, closer inspection of the rocks revealed that the slabs broken off the outcrop were hardly rocks at all, but a cemented amalgam of seashells and the branching skeletons of ancient ocean critters. It looked as though someone had taken a pickax to a coral reef. Here, there literally were no rocks that *weren't* fossils. We were 50 feet deep on the bottom of the sea, south of the equator, 450 million years ago. The rocks told the story of an alien planet that had unsettlingly little to do with the world above. Mouth agape, I suddenly understood the peculiar obsessions of the Dry Dredgers.

The Ordovician world is also known as the "Sea Without Fish." (As an indication of how much can change, the next major mass extinction would strike during the so-called age of fishes.) But

* Google it and prepare to be underwhelmed.

there *were* fish, even in the Ordovician. These were our ancestors: unimportant, small, and strange-looking—a group of largely jawless wallflowers next to the oceans' top predators. These rulers of the Ordovician were monsters without backbones, "creeping things that creepeth"—a swarm of shells, antennas, and tentacles.

To get to a mass extinction, you first need victims. Walking along the side of the highway outside Cincinnati (just past the Subway, Sprint, and Advance Auto Parts stores) is as good a place to start as any to meet this world eventually whisked away by the planet's first global slaughter of animal life. Spying an unusual rock, I brushed aside some plastic liquor bottles and pulled it from the rubble. The fossilized creature was curled into a ball, frozen with fear, and frozen for all time in stone.

"*Flexicalymene meeki,*" Dry Dredger board member Bill Heimbrock told me as I held it up to the sun.

"There's no flaws with it," he said. "It's perfect."

Aping the words I'd heard the veteran rockhounds using around me that day, I thoughtfully nodded and proclaimed, "Amazing preservation."

A few Dry Dredgers grumbled about my beginner's luck.

It was a trilobite, that staple of the natural history diorama, and part of a group of creatures that would endure a life-threatening blow here at the end of the Ordovician. Vaguely resembling the love child of an accordion and a horseshoe crab (the living animal to which they're most closely related), trilobites all but serve as the mascot of the Paleozoic era,* much as dinosaurs do for the Mesozoic, and mammals do for the Cenozoic. The trilobite is a mis-

* The age of animals has long been divided into three eras: Paleozoic, Mesozoic, and Cenozoic. The Mesozoic is broadly—though antiquatedly—thought of as the age of reptiles and the Cenozoic as the age of mammals. The Paleozoic comprises all the periods of animal life that came before the Mesozoic, including the Cambrian, Ordovician, Silurian, Devonian, Carboniferous, and Permian periods.

understood creature. The stereotype is that of a benthic Roomba mindlessly scouring the seafloor for hundreds of millions of years. And there were plenty of such unexciting, bottom-dwelling trilobites that shuffled between the horn corals and sponges of the ocean floor. But in the Ordovician there were also free-swimming trilobites, gliding through the open seas. Some sported the ultimate bug-eyes, dwarfing the rest of their bodies, while others were shaped like hourglasses, and still others like torpedoes. Some defy easy description: *ampyx*, for instance, had head shields decorated by long spikes, pointing forward and aft.* There were even large, free-swimming carnivorous trilobites in the Ordovician with streamlined heads that have been described as bearing a resemblance to "some modern, small sharks." Other mass extinctions might have had more charismatic victims, but just as the End-Cretaceous had *Tyrannosaurus rex*, the most fearsome dinosaur ever, watch an asteroid crash to earth, the Ordovician had *Isotelus rex,* the largest trilobite ever, as a witness to doomsday. At just under 3 feet long, this "giant" admittedly inspires less than mortal terror, but it was gigantic by trilobite standards.† The formidable *Isotelus rex* didn't survive the End-Ordovician mass extinction. Not much did.

"What was it afraid of?" I asked about my panicked fossil.

"Cephalopods," Heimbrock said, ominously. "Eurypterids."

It's a shame these animals don't have better names. Eurypterids are also known as "sea scorpions," and some were enormous,

* Other trilobites reveal more about their namers: Gregory Edgecombe of the London Natural History Museum named five species in the genus Articalymene, one for each member of the Sex Pistols, including *A. rotteni* and *A. viciousi*. The genus Mackenziurus has four members of the Ramones among its ranks, including *M. joeyi*, *M. johnnyi*, *M. deedeei*, and *M. ceejayi*—again thanks to Edgecombe.

† It was discovered in 2003 cemented in the rocky shores of Canada's giant Hudson Bay—ironically, one of the only extant Ordovician-style seas that sits on top of continental crust today.

with streamlined exoskeletons and carapaces housing a bouquet of dangling sci-fi appendages. In 2015, scientists working the Ordovician seas of Iowa found one such buglike beast the size of a human.

As for the cephalopods, a few feet away from my trilobite was the chambered cone shell of one of these animals—one that might have sent my fossil into its eternal death pose. Today cephalopods broadly include octopuses, squid, cuttlefish, and nautiluses (which can trace their lineage back to the Ordovician). Before the Ordovician, they grew to only a couple of inches at most, but by now they included astonishing animals like *Cameroceras,* which was housed in a cone shell that stretched almost 20 feet long. Museum reconstructions of the animal look something like an octopus jammed into a bus-sized ice cream cone. But to the trilobites, there was nothing ridiculous about the presence of these dreadnoughts hovering inches above the seafloor, tentacles ever searching the muck. At their Ordovician peak, these top-dog nautiloids numbered nearly 300 species. But when the ax of extinction fell, they were more than decimated: the cataclysm wiped out 80 percent of their ranks.

We know modern cephalopods, like octopus and cuttlefish, to be frighteningly intelligent, albeit with an alien intellect that developed along a completely different trajectory than our own; their brains bear almost no resemblance to anything found on our side of the family tree. Despite being mollusks, in the same group as insentient creatures like oysters and clams (which require no ethical consideration before swallowing whole), today octopuses are observed using tools, acting passive-aggressively toward their aquarium handlers, and, most dubiously, picking World Cup soccer games. Perhaps these first large cephalopods among the reefs of the Paleozoic mark the first sparks of subjective awareness—the beginnings of consciousness. Perhaps all of physical reality had unfolded in the billions of years since creation with no one there to notice it—until life emerged in those strange shallow seas over Cincinnati and elsewhere. Of course, this is all wild speculation, but it's fun.

The trilobite and cephalopod fossils along the roadside were embedded in heaps of still more fossil seashells. The shells, I quickly discovered, were so common and ubiquitous that they were considered too mundane to collect. These were the brachiopods—marine worms totally unrelated to the scallops and clams they resemble, having gone to the trouble of evolving their shells all on their own. The creatures were in such remarkable shape that they appeared to have washed ashore only days earlier . . . except that we were in the middle of the continent in a breakdown lane next to a strip mall, the shells were made of stone, and they were more than 200 million years older than dinosaurs. They also looked more Gothic than any seashells I'd ever seen: their two halves jaggedly interlocking, like bear traps: others had a more pleasingly sleek, Art Nouveau aspect, like the awning of a Paris Métro station, and still others resembled a geisha's fan. Though there are more exciting animals in the Ordovician than brachiopods, there are none easier to find by the handful. They absolutely paved the seafloors of the old earth, but would be savagely culled by the mass extinction.

I pulled up another peculiar rock from the midwestern seafloor and showed it to a Dry Dredger. The man—sporting a thick grizzled beard and a bandanna and looking as if he'd have been more at home in a motorcycle gang—took the fossil from me and pulled out a hand lens.

"Oh, that's a *leaverite,*" he said gruffly.

"Is that good?" I asked.

"Leave 'er right there," he said, chucking it on the ground. He was more excited by a slab I found plastered with the imprints of what looked like little hacksaw blades.

"Graptolites," he said, eyes widening. The sawtooth doodads had been built by bizarre little animals that lived tethered together in a sort of pelagic group home. They might have moved through the ocean by rowing in unison. And they spanned the earth before being all but obliterated by the mass extinction.

This was the world of the Ordovician: an odd sea world filled with invertebrates that for the most part make up in alien charm for what they lack in *T. rex*'s supersized ostentation. The world these animals inhabited was in some ways a version of our own, but it was also one so transformed by the intervening eons as to be scarcely recognizable.

During the Ordovician, a vast tropical sea covered most of present-day North America, in most places perhaps not much more than ankle- or knee-deep. Wading into the water on a sandy tropical beach in Wisconsin, you could keep trudging across most of the continent with your head above water before the seafloor dropped into the deep somewhere around Texas. This vast shallow province has been grandly dubbed the Great American Carbonate Bank—a nationwide Bahamas. Sea levels were possibly the highest in the history of complex life, and the shallow seas that drowned the continents were jammed with life. Flooded North America was rotated clockwise almost 90 degrees, California and the entire West Coast simply didn't exist, and chunks of New England, the Canadian Maritimes, England, and Wales—recently divorced from Africa near the South Pole—were an island chain called Avalonia not unlike modern Japan. Avalonia was then far away from the rest of North America, across the doomed Iapetus Ocean, the ancestor of the Atlantic.

Cincinnati provides just one window into the sea world of the Ordovician. Similar outcrops exist on almost every continent; some trilobites have even been found atop Mount Everest. The entire death zone of the world's highest peak is littered not only with skeletons clothed in the gaudy fluorescent parkas of climbing seasons past but with much older fossils as well—those of Ordovician trilobites and sea lilies. The ancient sea life was thrust up to the highest spot on Earth by the geologically recent collision of India and Asia.

Closer to Cincinnati during the Ordovician, 300 miles to the south, a shotgun blast of volcanic islands was colliding with what would eventually become the eastern edge of North America. This pileup was giving birth to the Appalachians, which once towered as high as the Alps above the drowned continent. Meanwhile, Kazakhstan, Siberia, and North China drifted far out at sea as solitary, island rafts, much of them covered with shallow oceans of their own. Microcontinents and archipelagoes like these littered the seas. It should be clear by now that making out the contours of our own world by squinting at this primeval geography is nearly impossible.

If you're not yet fully disoriented, across the sea South America was upside down and contiguous with Africa, as well as with Australia, India, Arabia, and Antarctica. Together they made up a supercontinent called Gondwana that was drifting over the South Pole. In artists' reconstructions, the continents are often shown as jigsaw pieces fitting together. This isn't quite right. Gondwana was one solid continent that was only later blown to pieces by processes deep within the earth. But—as one geologist told me—like gun violence, STDs, and world wars, tectonic boundaries tend to break out at the same places.

While the continents were flooded by high seas, what dry land did exist—like that in tropical Canada, Greenland, and the

Antarctic wastes of the southern supercontinent—was a vista of barren rock about as inviting as the feed from NASA's Mars *Curiosity* rover. Here on the rugged and bare continents there were no droning insects, no footprints, no trees, no shrubs—nothing. Life on land was relegated to a few damp patches of liverwort hugging the shore. Farther inland was an endlessly bleak and dusty wasteland. This was so long ago that rivers didn't even meander yet—the rooting plants that would have held back their banks would not exist for tens of millions of years. The day was 20 hours long and the night sky was filled with unfamiliar constellations. Carbon dioxide was far more plentiful in the atmosphere than today, trapping heat and offsetting the slightly dimmer sun that hung palely in the sky—keeping much of the world balmy and mostly ice-free.

Today much of the planet's landmass is in the Northern Hemisphere, but in the Ordovician almost the entire top of the globe was a vast ocean. At the bottom of these endless open seas, oxygen was in short supply. Much of the living world was instead jammed onto the shallow seas of the continents and dominated by submarine creepy-crawlies. But as indicated, this world was doomed. In the closing moments of the Ordovician, 85 percent of life on Earth would be wiped out.

If the mass extinction at the end of the Ordovician was extreme, it capped what had been an almost equally extreme flourishing of good times—a 40-million-year florescence of life, unlike any before or since. This was the Great Ordovician Biodiversification Event, the biggest expansion of biodiversity in the planet's history. Within one span of only 10 million years, the number of species on the planet *tripled*. Reefs began to grow in tiers and complexity, larvae took to the surface waters to avoid the gauntlet of tentacles on the seafloor, and animals began to burrow deeper in the muck to avoid the menace of squidlike monsters and giant sea scorpions. When geologists want you to know that an unappreciated event in earth history is really important, they telegraph it with an appropriately grandiose title, capitalize it, and tack on a "Great" for good measure. But perhaps aware that the words "Ordovician Biodiversification" don't exactly inspire awe in the general public, some have even tried rebranding the Great Ordovician Biodiversification Event as "Diversity's Big Bang."

What fueled Diversity's Big Bang is the sort of cutting-edge question that churns out PhDs. Once again, oxygen might have had something to do with it. Though the seas were still stifling by modern standards, there are hints of a growing oxygenation throughout the period. This might have been a product of life itself, which increasingly buried carbon in the seafloor, perhaps in huge algae blooms. The flip side of burying organic carbon is boosting oxygen; throughout the history of life, increasing oxygen levels have repeatedly spurred epochal innovations and experiments, like animal life—or, as we'll see later, nightmarishly large bugs. In the Ordovician, however, life might have been making the world increasingly more accommodating for more life to blossom.

Then there were the many islands strewn across the globe in the Ordovician, whose isolated shallow seas served as incubators for diversity. There's a reason why evolution was first discovered on islands—in the Galápagos by Charles Darwin, and independently in the Malay archipelago by Alfred Russel Wallace: islands drive biodiversity by separating populations, allowing them to pursue their own evolutionary stories and ultimately create new species. Indeed, the configuration of the planet in the Ordovician—with island-continents strewn across the tropics and subtropics—might have served as a sort of global Galápagos.

Some have even speculated that Diversity's Big Bang is owed

to a great collision in outer space 470 million years ago. In the lonely stretches between Mars and Jupiter, a soundless catastrophe destroyed an asteroid more than 100 kilometers in size, sending shards of the wreck swirling around the solar system. It was the largest asteroid breakup in billions of years. For a few million years afterwards, the earth absorbed the scattershot fallout from this collision in a hail of meteorites. Meteors might be more famous in geology as agents of wanton destruction (as at the end of the age of dinosaurs), but a 2008 *Nature Geoscience* paper argued that this Ordovician barrage from smaller rocks might have actually been a *good* thing, electrifying biodiversity by disrupting staid communities, clearing up ecospace, and just generally shaking things up. The enormous sea scorpion found in Iowa was discovered living in the watery ruins of one such crater dating to around 470 million years ago. Other craters of similar vintage are found in Oklahoma, in Wisconsin, and on the Slate Islands in Lake Superior. Across the globe, meteorite materials in Sweden, Russia, and China all similarly date to around 470 million years ago. But again, these asteroids pelted the Ordovician during the *best* of times. Even today, most earthbound meteorites come from the swarm of debris created by this massive primeval collision. In fact, according to the *New Scientist*, the only confirmed case of anyone ever being struck by a meteorite, a boy in Uganda in 1992, was from a fragment of this Ordovician wreck.

Meteor strikes weren't the only blows visited upon this alien world during its golden years. There was some homegrown chaos in the Ordovician as well, long before the mass extinction. In southwestern Wisconsin, the rolling patchwork of dairy farms is interrupted by a gash in the land where Highway 151 plows through the bedrock. Here road builders' dynamite revealed a tiramisu of ancient rock that now towers above the highway.

I joined a group of geologists on a field trip to this Wisconsin road cut. We trudged down the side of the road buffeted by the turbulent *whoosh* of passing tractor-trailers, craning our necks up at the striped wall. At its base, a jumble of brachiopod shells spilled out into the weeds and mingled with Styrofoam cups. Farther up the wall, weeds gained a foothold along two thin bands that cut through the ocean rock. These were ancient layers of ash, transmuted to clay after an eternity in the ground. The ash was volcanic, from some of the largest volcanic explosions in the history of complex life.

Catastrophic volcanic eruptions that have occurred in recent human history—like Krakatoa or Vesuvius—were pathetic burps compared to the Ordovician blasts that covered the ancient world in ash. The fallout from these ancient eruptions, known as the Deicke and Millbrig ash beds, can be seen in Ordovician rocks from Oklahoma to Minnesota to Georgia, spanning some 500,000 square miles. The ash layers dramatically thicken toward the southeastern United States, indicating that the monstrous volcanoes lurked perhaps somewhere off the coast of South Carolina; there a line of furious islands were barreling toward the edge of North America, ravenously chewing up the ocean floor underneath and exploding as they marched along. Over the ages the cataclysmic ashfall became bentonite, a clay mined for use in oil drilling and laxatives. A geologist on our trip picked up a chunk of the clay from the roadside cliff, popped it in his mouth, and grimaced, explaining that bentonite could be recognized from its consistency, which was something like toothpaste—minus, apparently, the minty freshness.

On the other side of the ocean, similar volcanic ash beds are recorded throughout Europe. The mammoth volcanic blasts must have seemed apocalyptic locally; they would have been so powerful that they would even have been *heard* across the planet. But as far as the fossil record is concerned—and much to the surprise of paleontologists—these Ordovician mega-volcanoes had hardly any effect on life at all. Not only did they have no effect, they, in fact, detonated during the heyday of Diversity's Big Bang, about 10 million years *before* the mass extinction struck. Apparently the planet can withstand a tremendous punch or two with good humor. It must take something truly dreadful to knock it over. Eruptions would continue until the extinction—when, curiously, the ash beds in the fossil record peter out and the volcanoes went quiet. Until that quiescence, however, the Ordovician was one of the most explosive periods in Earth's history.

That the most stunning period of biodiversification in the history of life took place at the same time as the planet was being pelted by meteors and unleashing some of its most powerful volcanic explosions ever is a testament to the living world's resilience. This confluence might even indicate that a little disturbance is a good thing for life. But the end of the Ordovician proves that a *big* disturbance can be a very bad thing indeed. Life was cresting as never before in the Late Ordovician. Then it was suddenly poleaxed by extinction.

"Cincinnatian time was significant in the history of life as a Golden Age of evolutionary diversification just before a major crisis of mass extinction," writes geologist David L. Meyer. "Few if any fossil species found in the Cincinnatian strata survived."

The peculiar trilobites, the nautiloid cephalopods, the brachiopods, the graptolites—nothing I found on the side of that midwestern highway would escape the harrowing scythe of mass extinction.

So what happened?

The Dry Dredgers don't collect fossils from the end of the Ordovician. It's not some peculiar club policy. There are no ocean rocks to collect in Ohio from the very end of the Ordovician period, because the ocean suddenly drained away from the Midwest, leaving this shallow sea world gasping.

I met Seth Finnegan at the *Tyrannosaurus rex* skeleton in the UC Berkeley Paleontology Museum—possibly the most famous mascot for mass extinction. I was there to talk to him about an Armageddon that happened nearly 400 million years before *T. rex*'s ballistic encounter with the solar system. Finnegan and his colleagues have been slowly piecing together the story of the End-Ordovician mass extinction with lab equipment and computer programs in university offices, as well as with rock hammers and camping gear in the middle of nowhere.

Finnegan is enthusiastic, cuttingly intelligent, and almost compulsively funny. He drolly laments that the fossils of the eastern half of North America are mostly covered over by "photosynthetic glop" (i.e., plants and trees), and in the boozy after-hours of geology conferences, he regales colleagues with tales of amateur crackpots who show up at the doors of his department with eccentric, if dubious, fossil finds from greater San Francisco.

I had come to Berkeley to pick his brain about a talk I had heard him give at a conference in Vancouver earlier that year. Finnegan had chewed over the possible causes of the End-Ordovician extinction, which he had been teasing out of the rocks with the help of a variety of algorithms and machine-learning computer programs with names like "gradient boosting models" and "multinomial logistic regressions." Clearly, the stereotype of the paleontologist as a musty old scientist dusting off bones in some forgotten corner of a natural history museum needs updating.

At the Vancouver talk, Seth had rattled off the many causes that have been variously proposed for the End-Ordovician mass extinction, which wiped out up to 85 percent of animal life on earth almost half a billion years ago. He evaluated the many possible killers, hints of which lurk in the fossil record and in ancient rocks that span the globe. One suspected killer came in for his special derision.

"And then there's the gamma ray burst hypothesis!" Finnegan had said grandly. "Which I haven't heard about in a while, but it's still up on Wikipedia." The audience of geologists had erupted in a knowing laughter.

The sad truth is that almost no one outside of a small guild of invertebrate paleontologists (and some oil companies) cares at all about the Ordovician period. When journalists do deign to mention this 50-million-year stretch of earth history, it usually serves only as a proper noun obscure enough to lend rhetorical weight to an exciting pop-science idea that virtually no one in the paleontology community takes seriously.

From *Scientific American:* "[Gamma ray bursts] might have already hit Earth, at the end of the Ordovician period nearly 450 million years ago."

From *National Geographic:* "A brilliant burst of gamma rays may have caused a mass extinction event on Earth 440 million years ago."

Gamma ray bursts are the most powerful blasts of radiation known in the universe. They're thought to be generated when extremely large stars violently collapse into black holes, shooting death-ray jets of radiation out of their poles that—for just a few seconds—are visible across the universe. Obviously any planet in the path of one of these sterilizing blasts would be toast at short range, and the possibility of one hitting the planet is an undeniably sexy scenario for mass extinction. Besides, the left-field idea that an asteroid killed the dinosaurs was once considered heretical to paleontologists, so perhaps the idea that a crippling blast from beyond wiped out the Ordovician world—an idea first floated by astronomers at the University of Kansas in 2003—wasn't so crazy after all.

But determining whether a gamma ray burst actually hit our planet in the ancient past is nearly impossible. The theory does make at least one prediction: that the extinction would be much worse on the hemisphere of the planet facing the cosmic blast than it would be on the opposite side, shielded by the rest of the earth. Unfortunately for gamma ray burst proponents, there is no such Ordovician extinction signal, with fossil life on only one side of the globe being slaughtered. Unfortunately for life on Earth, the mass extinction was a truly global phenomenon.

A gamma ray burst would also inflict a geologically instantaneous razing of the biosphere, but the End-Ordovician die-off was carried out in two discrete pulses of extinction separated by hundreds of thousands of years. Nevertheless, for some reason, gamma ray bursts are mentioned in almost every account in the popular press that bothers to mention the End-Ordovician mass extinction. Each geologist and paleontologist I asked quickly dismissed the idea, but not before registering a mild annoyance at its zombielike persistence in the media.*

"There's no evidence for it whatsoever," said Finnegan.

While there's no evidence for a death ray from outer space, there's plenty of evidence for other cataclysms closer to home.

To understand this unthinkably remote chapter in Earth's past, and the bitter end of the Ordovician, we first need to take a brief

---

* University of Durham paleontologist David Harper told me that for one paper he wrote on the extinction, a reviewer even insisted that he delete a paragraph in which he *dismissed* the gamma ray burst hypothesis, protesting that even an unflattering reference to the idea in an academic paper brought it undue respectability.

detour hundreds of millions of years forward in time, to our own geological yesterday. Not very long ago, the Northern Hemisphere was smothered by ice and the sea level was 400 feet lower than it is today. Right now, far offshore at the bottom of the Atlantic Ocean, sea robins and cod tend to mastodon and woolly mammoth graveyards. Their tusks are pulled up by scallop dredgers on George's Bank and in the Gulf of Maine. Though they're found at the bottom of the ocean, these were not amphibious mammoths. Instead, they roamed a vast coastal plain on what was dry Atlantic continental shelf before the great ice sheets melted and raised the seas hundreds of feet. Underwater canyons now teeming with sea life were rivers and scenic estuaries cutting through the dry seabed, and all this was only a couple of hundred human generations ago—essentially *now* from a geologic standpoint.

Of the 4,500 million years of the earth's existence, the most recent 2.6 million have been a relatively atypical age of ice—huge stores of the earth's water have been locked up in polar ice caps and ice sheets. This is the ice age of popular imagination—the one they make animated kids' movies about. But it wasn't the first, or only, ice age in the history of the planet.

Surprisingly, our ice age—which once hosted woolly mammoths and saber-toothed cats—isn't over; it's just on recess. Throughout the ice age of the past few million years, there have been dozens of so-called interglacials—brief windows of warmth, lasting only a few thousand years, when it gets warmer, the ice rapidly melts and retreats to the poles (where it is today), and the seas rise by hundreds of feet. We're currently in one of these brief respites from the cold, but interglacials don't usually last very long. This is all caused by the periodic wobble of the planet in space and the rhythmic changes to its orbit, which nudges the planet in and out of the sunlight, alternately locking much of the Northern Hemisphere in ice and then thawing it over and over again, like a geological metronome. The thaws often last fewer than 10,000 years before the great ice sheets at the poles begin to advance on the continents once more, sending the seas plummeting hundreds of feet.

There have been at least twenty such balmy intermissions like our own sprinkled throughout the past few million years of our ice age. But unlike the many previous warm interglacials, civilization—and all of recorded human history—happened to arise during this one. Our few millennia in the sunshine are up, and if it weren't for us, we might be just about ready to leave this agreeable little interregnum and jump back into the ongoing deep freeze of the Pleistocene for 100,000 bitterly cold years. (Obviously, because of the ways humans have profoundly transformed the chemistry of the earth's oceans and atmosphere in only the past few decades, this regular schedule has probably been upended and it's not about to get colder anytime soon.)

So how do we recognize ice ages in the rock record? Well, we know about our own (very) recent and ongoing cosmic freeze-thaw cycle from (among other things) the study of fossil foraminifera—tiny plankton that record the signatures of the climate in their shells in the form of oxygen isotopes. Over thousands and millions of years, they fall dreamily through the oceans and blanket the sea bottom in a perpetual snowfall, inviting scientists to drill cores into them for chemical clues to a bygone planet. But it doesn't take isotopic analyses or drill cores to reveal that the world we take for granted was, until very recently, frozen solid. As I write this in Massachusetts, the evidence surrounds me, plain to see. Massive boulders unceremoniously dropped by the glaciers dot the deep woods, town centers, and beaches throughout New England. Kettle ponds mark the spots where large solitary hunks of ice, orphaned by the great ice sheets, were left behind to melt in their tracks. The winter world is evident in the grooved lines, called striations, etched in the bedrock of the mountains of New Hampshire, where mile-thick grindstones of ice advanced and retreated, scraping bare the scenery behind them. Long Island, Block Island, Cape Cod, Martha's Vineyard, and Nantucket are all essentially dump heaps of rocks and sand where the ice sheets pushed south, sputtered and spewed their guts out onto the tundra. Like the many unremembered islands and sand spits that came and passed in the interglacials before, these features are all being quickly eroded away to nothing.

These topographic ghost stories of a planet recently locked in ice are told far beyond my own neighborhood: the Finger Lakes of New York were carved by massive glaciers, while the Great Lakes are basically the world's largest puddles, left when the ice sheets melted only a few thousand years ago. The most dramatic example might be in the epic Channeled Scablands of eastern Washington, which were carved by truly mind-blowing cyclical floods called *jökulhlaups*.

During the last glacial cycle, as a massive ice sheet pushed into Idaho, the ice blocked the Clark Fork River and created an enormous dammed lake in Montana, six times the volume of Lake Erie. As the lake grew and grew, it eventually reached a depth of 600 feet, at which point the obstructing ice started to float. As water continued to gnaw at cracks at the base of the ice dam, the lake suddenly failed and the whole system collapsed in a catastrophic flood that—all at once—released ten times the flow of all the rivers in the world. Waves leaving ripple marks 450 feet long and carrying 30-foot boulders crashed through eastern Washington, tearing up the bedrock, sculpting canyons, and leaving the southeast corner of the state barren of soil and vegetation. When the glacial dam began to re-form, the lake began to fill once more, and eventually it burst again. And again. This catastrophe repeated itself perhaps as many as sixty times between 15,300 and 12,000 years ago. This was right about when the first Americans were wandering into the continent, and perhaps some unlucky few witnessed, and even perished, in these local apocalypses. Mammoths and other animals certainly did. Their bones are found in the jökulhlaup flood deposits. Looking for similar deposits in much older rocks can tell us when even more ancient ice ages befell the planet.

But *our* ice age was so recent that some places like Alaska and Canada are literally still bouncing back up from the removal of oceans of ice overhead—that is, the landmass is actually rising, year after year, like your seat cushion after you stand up. The height of this last ice age is popularly conceived to be a distant part of the planet's past. But from a geological perspective, it was an eyeblink ago. If all of Earth's history were represented by a 24-hour clock, it was half a second before midnight.*

Far from the kettle ponds and erratic boulders of New England, however, in the center of the Sahara Desert, a lonely outcrop of rock sits exposed to the searing afternoon sun. It has been laid bare by the harsh excavation of a passing sandstorm. Curiously, the shifting sands reveal that this barren rock has the same striations across its surface as the grooved stones of New Hampshire and Maine, as if scratched into the bedrock by the fingernails of giants. From Mauritania to Saudi Arabia, sun-baked and scarred rocks testify to a landscape once pulverized by ice. In the Anti-Atlas Mountains of Morocco there are drumlins sculpted by ice streams and enormous tunnel valleys carved by glacial meltwater. In Libya, one of the hottest countries on Earth, geologists find still more of these glacial tunnel valleys, and on its border with neighboring Algeria there's even evidence for cataclysmic jökulhlaups. Rocks once borne by rafts of ice have been found dropped in Ethiopia and Eritrea, and glacial sandstones and till are found throughout the scorching sand seas of the Sahara and Saudi Arabia. This land-

* Viewed from the dwarf planet Sedna, which roams the lonely outer reaches of our solar system, it's only been a little over year since the ice age.

scape was not molested, however, by the glaciers of the past few million years. Instead, these are the vestiges of a truly ancient ice age. The ice sheets that vandalized this desolate desert rock scoured the earth, not a few thousand years ago, but some 445 *million* years ago. These rocks mark the end of the Ordovician.

The evidence for a massive glaciation at the end of the Ordovician is surprising. Until this frosty climax (it's long been thought), this was a warm world, with atmospheric carbon dioxide perhaps eight times higher than today. More recent evidence, though, points to a planet that was, in fact, cooling in the final few million years of the Ordovician. But at the bitter end, glaciers suddenly swelled on Antarctic Africa and stole their water from the ocean, dropping the sea level by more than 300 feet. This explains the disappearance of the fossil record in the shallow seas of Cincinnati and may well go a long way toward explaining the extinction itself. For a world that lived mostly in these shallow continental oceans, such a sudden and dramatic drop in the sea level seems like it would have been apocalyptic. For places like Cincinnati and much of the rest of the submerged continent, the seas were literally drained away, leaving the former seafloor and all of its inhabitants out to dry under the Ordovician sun for a million years. Nautiloids and *Isotelus rex* alike found their vast, shallow continental playgrounds reduced to thousand-mile tracts of limestone ruins, now crumbling and exposed to the wind.

This is the rough sketch of the fallen world at the end of the Ordovician, but Finnegan's team wanted to know the grisly details. In particular, they wanted to know how much the climate had to change to bring on this icy apocalypse. Rocks leading up to the extinction aren't hard to find, but with the fossil record almost disappearing from the continents right at the extinction—and the ocean floor further offshore having long since been chewed up and destroyed in subduction zones*—researchers have had to be resourceful in getting data on the disaster. They have had to seek out rare spots on Earth that, through the quirks of tectonics, managed to both stay underwater, gathering a fossil record through the sea level drop, and avoid getting destroyed or defaced in the tectonic train wreck of the ages to come.

One such place is Anticosti Island in Quebec. "It's one of the largest islands you've never heard of," said Finnegan. "It has a population of like 250 people, 150,000 deer, and a functionally infinite population of mosquitoes and blackflies—so it's a fun place to work."

Far from the Bay Area, guarding the mouth of the St. Lawrence River in Quebec, Anticosti Island, like much of the North, is still rising up after having been just released from the mile-thick ice sheets of our own ice age. In the island's slow ascent from the Gulf of St. Lawrence, stunning white cliffs have surged from the sea, spilling out 445-million-year-old coral reefs that span the great Ordovician extinction. Though rebounding from recent glaciers, these cliffs provide insight into the far more ancient ice age that hammered life at the end of the Ordovician. Inland, rivers carve out still more cross sections of these ancient tropics in narrow, fossil-rich

---

* The rocks formed far offshore at the bottom of the deep ocean might provide more insight into the crisis, but they have long since been destroyed, owing to the fundamental difference between continental crust (like that beneath Cincinnati) and denser offshore ocean crust. Continental crust is less dense and floats atop Earth's mantle like froth on a boiling pot of water, enduring practically forever, while denser offshore ocean crust is continually created along spreading midocean ridges and mostly destroyed in ravenous subduction zones, where it's thrust back down into the earth again. As a result, the oldest parts of the ocean floor today are less than 200 million years old, created when dinosaurs were trampling around during the Jurassic. In other words, the rocks of the modern ocean floor are hundreds of millions of years too young to tell us anything about the Ordovician.

canyons that slice through the boreal wilderness.* Consequently, the island is also of interest to the oil industry: this same Ordovician sea life that buried carbon over the ages has been transmuted into fossil fuels. They're called that for a reason.

"These are some coral colonies, and here are some brachiopods," said Finnegan as he pulled up pictures in his Berkeley office of the reefs cemented in the Québécois cliffs. The reefs were made of strange, long-extinct corals that resembled horns of plenty reaching toward the sun. "And these are really weird," he said. "This thing that looks like a logjam, these are big calcifying sponges that grew vertically like trees on the seafloor."

Hacking chunks off of these reefs and bringing them back to the lab—where they subjected them to a bit of geochemical wizardry†—Finnegan's team discovered an abrupt drop of about 5 degrees Celsius in the tropical ocean at the end of the Ordovician. Five degrees might not sound like a mass extinction, but the rocks say otherwise.

"The general consensus in the field is that the Ordovician mass extinction is closely tied to climate change," he said.

This cooling is in keeping with what's seen in the fossil record: a wholesale razing of tropical sea life and a temporary takeover by creatures from the poles. And it lines up nicely with the intense glacial features that are so striking in the landscapes of the Sahara.‡

---

* Aka photosynthetic glop.

† Specifically, carbonate clumped isotope paleothermometry.

‡ In fact, the Saharan glaciation is even apparent in Finnegan's data from Quebec. Oxygen isotopes in the ancient reefs of Anticosti Island suddenly become unusually heavy at the end of the Ordovician. That means that, at least in the tropical ocean, a lot of lighter oxygen isotopes had gone missing. Because lighter isotopes are literally lighter, they evaporate more easily from the sea, and this isotopically lighter water wandered over Africa and fell out as snow, forming the vast ice sheets that would come to cover Gondwana. As a result, the seawater left behind in the oceans was isotopically heavier, as were the corals and seashells that grew from them. To account for the huge shifts to heavier values found in these tropical reefs at the end of the Ordovician,

The question then is this: why do bad ice ages happen to good planets?

"So carbon dioxide is what controls climate, right?" said Harvard geologist Francis MacDonald. He was being flip, of course: lots of things control the climate, like solar intensity, surface reflectivity, and ocean circulation, to name a few. But, *ceteris paribus,* as carbon dioxide goes, so goes the climate. This, it should be noted, has been an uncontroversial tenet of geoscience for more than a century. The greenhouse effect was first described in the 1820s by French physicist Joseph Fourier, who noted correctly that the planet would be uninhabitably cold if it weren't for Earth's blanket of insulating gases. In 1859, Irish physicist John Tyndall discovered carbon dioxide to be one such greenhouse gas, and in 1896 Swedish scientist Svante Arrhenius predicted that doubling CO_2 in the atmosphere would warm the planet by about 4 degrees Celsius, a prediction that's roughly in line with those of our most powerful modern supercomputers. Needless to say, discussion of this basic science by actors with baldly political motivations can be depressing beyond words.

I met MacDonald in his Cambridge office, which isn't far from where he does much of the fieldwork for his study of the creation of the Appalachian Mountains: the backyard of New England. "So what I'm trying to explore is the way in which some of the large-scale tectonic change we see in earth history is related to the environmental change we see on the surface. . . . So maybe we're actually taking more CO_2 out [of the atmosphere] at some times than at others. So why would that be?"

Today the concern is about injecting carbon dioxide into the atmosphere too quickly and creating a global hothouse climate. But

---

Finnegan determined that the ice sheets that suddenly developed on the other side of the world must have been enormous—substantially larger than even those seen during the most punishing ice ages of our recent geological past.

just as problematic can be quickly plummeting levels of carbon dioxide, which can create an icehouse climate instead. However strange it seems at first, the creation of the Appalachians might hold the key to explaining this punishing glaciation that nearly wiped out life on Earth.

To understand much of what follows in this book, and much of earth history, we need to make another brief (and hopefully painless) detour into geochemistry.

Carbon dioxide reacts with rain to make it slightly acidic. This slightly acidic rain lashes rocks over millions of years, breaking them down and washing stuff like calcium out into the rivers and, eventually, into the sea. This carbon- and calcium-rich broth is then incorporated into the bodies of living things like sponges, corals, and plankton. These creatures then bury the carbon at the bottom of the ocean as calcium carbonate limestones. Visit your favorite monument or building made of limestone and take a closer look—the stone is simply the detritus of living things.* This is how carbon dioxide from the atmosphere is converted to rock and safely stored away in the earth. Eventually this process might dangerously drain away the atmosphere's blanket of carbon dioxide that keeps the planet habitably warm. But this *doesn't* happen because carbon dioxide is slowly but constantly being replenished elsewhere on Earth by emissions from volcanoes on land and at mid-ocean ridges. But today humans—by retrieving and burning hundreds of millions of years' worth of this carbon buried by geology—contribute 100 times more carbon to the atmosphere every year than volcanoes. In the Ordovician, profoundly ill-tempered volcanic island chains—like the ones that produced the

* Unless it's made of travertine.

ashfall I saw in Wisconsin—played the role of the modern power plant, contributing vast amounts of CO_2 to the atmosphere and keeping the planet warm.

The earth has a brilliant way of dealing with too much CO_2. When carbon dioxide in the atmosphere goes way, way up from increased volcanic activity (or from, say, coal-fired power plants), the planet warms up from the greenhouse effect.* But the catch is that in this warmer, stormier, high-CO_2 world, CO_2 is also drawn back down into the earth even faster. This is because the more acidic rain, warmer temperatures, and increased rainfall created by excess CO_2 all work in concert to intensify rock weathering. This causes the planet to cool off more quickly when it gets too warm, by drawing down more CO_2, which ends up as limestone in the ocean. When the planet finally cools, the processes of rock weathering slow down as well, the drawdown of CO_2 relents, and the planet returns to an equilibrium.†

This is the carbonate-silicate cycle. It's our planet's improbably effective way of regulating the climate. It's also known as "Earth's thermostat." But sometimes the thermostat breaks.

"People say if it's warmer we'll have more weathering, and if it's colder we'll get less weathering," MacDonald said. "In that case, we should have an equable climate throughout all of earth history. Well, sorry, it failed catastrophically with Snowball Earth. But it also failed at the end of the Ordovician. So why is this failing?"

One way to draw down lots of carbon dioxide quickly and break this planetary thermostat is to suddenly thrust up an epic, volcanic mountain chain thousands of miles long in the middle of

---

* Again, it has to be emphasized, this is an uncontroversial idea in the earth sciences since at least the American Civil War.

† Unfortunately for humans, it will take about 100,000 years for these weathering processes to remove anthropogenic carbon dioxide from the atmosphere.

the tropics, shoving more rock up where it's warm and wet and weathering is at its most intense.

"It's a matter of creating fresh surfaces to weather," said MacDonald. "So one good way to create fresh surfaces is to actually lift up a mountain somehow and keep shedding it and eroding it."

The modern Appalachians began to form in the Ordovician* when a volcanic island chain with an insatiable appetite for ocean crust ate its way across the sea before plowing into the eastern edge of North America. This train wreck is visible throughout the mangled rocks of New England. At Arrowhead—the farmhouse in Pittsfield, Massachusetts, where Herman Melville wrote *Moby-Dick*—the author drew inspiration for his white whale by gazing out his desk window at the snow-covered hump of Mount Greylock, which rises above the surrounding hills like a whale breaching the wine-dark waves. Surprisingly, Melville wasn't far off with this nautical inspiration. The mountain, the tallest in Massachusetts, is actually made of Ordovician seafloor; it was thrust up onto the continent by an approaching line of volcanic islands and the ocean plate they rode in on.† The roots of these volcanic islands are visible as outcrops of Ordovician gneiss throughout New England, perhaps nowhere more spectacularly than under the New England Power Company dam in downtown Shelburne Falls, Massachusetts, where a swirling expanse of 475-million-year-old magma has been newly pitted with gigantic "potholes," gouged into the rock only a few thousand years ago by waterfalls cascading off melting ice age glaciers. Equipped with the decrypting lens of geology, the roadside scenery of New England reveals this monumental collision that created the ancient Appalachians

---

* Parts of the Appalachians reveal an even older ancestral mountain chain formed by continental collisions that created an ancient supercontinent more than a billion years ago.

† Mount Greylock is also covered in glacial striations from the most recent ice age.

in the smashed confusion of rocks that marbles the sides of the highways. The east coast of North America isn't the familiar extension of the continent it appears to be, but is instead a smattering of old islands and volcanoes that have been grafted onto the side of the continent and crumpled in spectacular collisions.

The mountains produced by these tectonic train wrecks in the Ordovician might have reached as high into the heavens as the Himalayas and stretched from Greenland to Alabama.

Surprisingly, this venerable mountain belt might have played the role of prime mover in the mass extinction. As the Appalachians climbed toward the heavens, fresh, weatherable volcanic rock was being constantly pushed up to the skies to be eroded away, pulling down the atmosphere's carbon dioxide with it.

"So this should be a pretty good recipe for a lot more silicate weathering and sucking up CO_2," said MacDonald. "And these volcanic rocks are a lot easier to break down than, say, old continental rock."

This carbon from the atmosphere and the minerals washing out of the eroding mountains were fueling the explosion of Ordovician life in the shallow seas over the Midwest and being buried as limestone in places like Cincinnati in the bodies of animals. If MacDonald is right, the creation of the Appalachians is responsible for a precipitous drawdown in carbon dioxide and the brief ice age that nearly wiped clean the evolutionary slate 445 million years ago.

It's an intuitively satisfying story, but is there any evidence that the CO_2-swallowing processes of rock weathering did in fact intensify in the run-up to the Ordovician glaciation and mass extinction? Or even that episodes of intense rock weathering can lead to an ice age? There is. Strontium isotopes in the rock record can be examined to track periods in Earth's history when weathering has been particularly intense, and for our own ice age, as the modern planet has cooled from the greenhouse of the dinosaurs to the icehouse of the mammoths, the strontium isotope record goes haywire right as India first smashes into Asia, pushing the Himalayas up to the heavens (along with their Ordovician fossils) to be weathered away.

"That [collision] happened right around the very first initiation of the Antarctic ice sheets," said MacDonald. "So I just find that coincidence so compelling that I can't ignore it. Okay, so before that, what's the next big long mountain chain that's happening at the equator? And I have to go back to the Ordovician."

MacDonald's colleague at Ohio State, Matthew Saltzman, went searching for a similarly telling strontium signal in Ordovician rocks. And in remote corners of the Appalachians, and in Nevada, he found it.

"Something like 465 million years ago, the strontium isotope ratios make an unambiguous plunge," Saltzman told me in his office in Columbus. "And the simplest interpretation of that—which is always what we're looking for—is that you're weathering fresh young volcanic rock. So you're left with the conclusion that that would have drawn down CO_2 levels in the atmosphere. But again, we're about 20 million years before the extinction itself. . . . You do start cooling down, but there's obviously some thresholds involved to get you from a climate that's cooling to one that becomes an icehouse."

In other words, forming ice sheets is not a linear process. They don't grow on the continents like mold on bread, but instead explode in size once some climatic tipping point has been crossed. For the modern world, that tipping point was reached 2.6 million years ago, when the planet finally crossed the glacial point of no return and the Northern Hemisphere joined Antarctica in an icy alliance. For the Ordovician, the tipping point wasn't reached until the very end of the period, 445 million years ago, and it nearly cost the planet its life. The deep freeze might have been delayed until then by a continuous outpouring of CO_2 from the very sorts of volcanic eruptions I saw layered in the rocks of Wisconsin, which kept the climate balanced on a knife's edge until the bitter end.

"So maybe some of the same volcanism that was producing these basalts to be weathered was somehow moderating the climate just enough so that it didn't get very cold until the very end," said Saltzman. "You've got these huge ash beds going up through the Ordovician and then basically ending. So as you'd expect, you have a lot of volcanism, which would have a warming effect [from the CO_2] at the same time you've got this pull in the other direction of CO_2 being drawn down from weathering."

When the volcanoes finally went quiet at the end of the Ordovician, this steady supply of CO_2 to the atmosphere was shut off. But the weathering of the volcanic rock continued apace, and the CO_2 in the atmosphere began to plummet.

So where did the carbon from the atmosphere go? A lot of it ended up in life itself. As the nutrients from the battered rocks were washing off the mountains and into the seas, they fueled blooms of plankton, which sank to the bottom of the sea, burying the carbon from the atmosphere in them as they did. These blooms of ancient sea life have been good news for oil and gas companies that prospect Ordovician rocks, but unfortunately for life, this carbon burial might have hurled the period toward its icy crescendo.

"This burial of petroleum source rocks was cooling things down," said Saltzman. "And it's all related to the mountain building because, on the organic side of things, once you start weathering these rocks and bringing in nutrients like phosphorus from the land, you would expect there to be positive feedbacks towards cooling."

Finnegan summed up Saltzman's findings: "With respect to CO_2, volcanism giveth and it taketh away. You get a lot of CO_2 outgassing while it's going on, but you create a bunch of fresh volcanic rock [that] you then weather and draw down CO_2."

So there you have the three ingredients of a mass extinction. A world that lived in the shallows; a CO_2-sucking mega-mountain chain that chilled the planet to a deep freeze; and a supercontinent straddling the South Pole, providing a convenient place to put all that ice. But that's not the whole story.

Unlike our ice age, which began in earnest 2.6 million years ago and tentatively retreated only a few thousand years ago, the glaciation at the end of the Ordovician 445 million years ago killed almost everything on the planet. The most curious part of the die-off, and the nagging detail that pushes it into true mass extinction territory, is that it wasn't just the creatures left high and dry on the continents, like our friends in Cincinnati, that were wiped out. Animals that swam out in the open seas and those that lived in the deep fared no better. This wrinkle in the story means that there must have been more to the extinction than the (somewhat) simple version outlined so far—of emerging ice and disappearing seas. Why was this brief Ordovician ice age apocalyptic but our modern ice age has had comparatively little effect on life at all (until the recent spread of humans)? This is a question that still perplexes paleontologists.

"You can't just focus on the nature of the change," said Finnegan. "You need to consider, 'What's the starting state?' And the starting state is really different in the late Ordovician. It really is a different world."

We take for granted the shape of our world and the position of the continents—the familiar geography that seems as eternal as the order of the planets.* But this arrangement is temporary: it isn't how the planet has been and it isn't how it will be. This ever-shifting world map has major implications, far beyond cartography. The accidental orientation of the continents has a profound influence on life. A few million years ago, when the planet descended into its current ice age after its long slow decline from the greenhouse climate of the dinosaurs, the world was configured in a very peculiar way: namely, the way it's configured now, with long north-south coastlines that extend from the tropics nearly to the poles. For example, the turnip-like South America and its North American thought bubble stand upright and span almost every latitude. This arrangement has been a lucky one for the animals trying to navigate a finicky climate that, in recent history, keeps plunging in and out of ice ages. When the cold comes, or when the warmth returns, most of them have simply been able to march up and down the continents to stay comfortable.

"So if you look at a modern world, when we finally descended into deep glaciation, the world is composed of these long north-south linear coastlines and it's fairly easy to shift your range and track your preferred climate," Finnegan said. "And we see evidence that that's exactly what happened."

He brought me into his lab and hauled out a plastic bag filled with jumbled fragments of abalone, limpet, and scallop shells. Still iridescent, they looked as though they might have washed up on the beach during the last storm. In fact, they were 130,000 years old, from the last warm interglacial, when California's climate was

* Which, interestingly, is also subject to change.

similar to today's. The shells were gathered at San Nicolas Island off Los Angeles. When the ice retreated, these mollusks simply marched hundreds of miles up the coast from Central America. When the interglacial was over and the ice returned, they wandered back down to the tropics, tracking the dramatic changes of our ice age. Whether plants and animals have these escape routes available when a new wind blows can be the difference between survival and extinction.

Today plants and animals are already shifting their ranges in response to man-made changes in climate. In 2012 the US Department of Agriculture was forced to update its vegetation maps to reflect the northward shift in plant life under way in the United States. Lobster fishermen south of Massachusetts have all but shuttered their businesses as the crustaceans steadily march northward to track the cooler bottom waters they prefer. Zooplankton in the North Sea has shifted its range toward the poles by 700 miles in the past few decades, causing chaos for fisheries managers, who have had to oversee a suite of new southern species entering their jurisdiction. And this great march north is just beginning.

In extremely warm, extremely high-CO_2 periods in Earth's history, time-traveling tourists might be taken aback by some of the more bizarre sights, like crocodiles sunbathing in the Arctic Circle 55 million years ago. But in the coming decades, these animals might find it more difficult to trudge north to these refuges once more. Unlike their itinerant ancestors, when modern crocs start their poleward march, they'll push into highly developed coastlines and drained wetlands inhabited by hotels, golf courses, summer homes, and coastal residents unlikely to welcome them to their shores. But the animals of the Ordovician 445 million years ago had a problem that was worse than man-made obstacles to migration: migrating in the face of the End-Ordovician climate convulsions might have been simply impossible.

It's been argued that the many island continents of the Ordovician might have been a driver of Diversity's Big Bang, but Finnegan thinks that this disparate island planet had its fatal downside as well. When the climate flipped, the animals found themselves marooned on their island continents, separated from habitable refuges by vast open oceans.

"So geography may have played a role in the diversification, but I would also argue it may have also played a role in the extinction," he said. "The question really is: is there anywhere they can go to if the climate changes that's still going to be within the climate window they were adapted to? If you're on island continents, that's a little harder to do."

When the world changed at the end of the Ordovician, there was simply nowhere to escape to.

Certain features of the hoary Ordovician extinction lend themselves to modern comparisons—the role of carbon dioxide–driven climate change and habitat destruction, for instance. But there are certain aspects of the ancient extinction that strain any analogy to our modern world. The Ordovician was, after all, almost a half-billion years ago on a planet that, even if viewed from space, would be completely unrecognizable to us. In many ways the planet operated completely differently. One of the more unusual aspects of the extinction is that many animals that lived deeper in the ocean might have been, counterintuitively, wiped out by an *increase* in oxygen. After all, this was a time when oxygen in the oceans, though building, was still quite low. Making peace with these stifling conditions could be a prudent strategy. One animal that nearly perfected this strategy was the brachiopod.

Comparisons to the charismatic fauna that would come to characterize the later stages of life on earth can make Ordovician researchers defensive about the unlovable objects of their affection. When the listless brachiopod is compared to a dinosaur, an animal whose blockbuster appeal is immediately apparent, the retort is quick from those who study marine invertebrates of the sort that mucked about in the prehistoric seas: anyone can love a dinosaur, but it takes a true diehard to appreciate the life—if one can call it that—of a brachiopod.

"You have to understand," Finnegan said half jokingly, "most paleontologists regard the dinosaur folks the same way marine biologists look at people who work with dolphins."

Amazingly, there are still a few brachiopods kicking around the planet today, resigned to relive their Paleozoic glory years in rare refuges, like around New Zealand and in Southeast Asia, where (once smothered in spices) they make a rubbery, anachronistic delicacy. Researchers who have brought these relics into the lab to probe them for insights into life in the Paleozoic have found the animals to be among their more uninspiring subjects.

"They don't do much," said Finnegan.

At a seminar for grad students I attended at the University of Chicago, Stanford paleontologist Jonathan Payne described the challenges of studying the torpid creatures, which make clams and mussels look like white-hot metabolic furnaces by comparison. If you're feeling cruel, you could lock a brachiopod in a jar of water and prepare to wait weeks before the animal finally succumbed to a pathetically slow death by suffocation. When one dies in the lab, it's nearly impossible to distinguish it from a living specimen. Often the only indication is that they start to smell. As Payne explained, when scientists want to measure the metabolism of an animal they typically stick it in an aquarium and measure the amount of oxygen it draws down.

"For brachiopods, this turns out to be hard because they have such low metabolic rates that when you stick them in a tank, you actually have to spike the tank with strong antibiotics to make sure that you're not measuring the background respiration of the bacteria in the tank," he said.

"To a first approximation," paleontologist Susan Kidwell chimed in, "they're dead." The audience of aspiring paleontologists roared with laughter.

"When I'm teaching, I often joke that brachiopods are sediment, they're not really organisms," added Payne, pouring it on. But Payne wasn't just workshopping brachiopod stand-up. The response of brachiopods to the upheavals at the end of the Ordovician has helped scientists parse exactly what was going wrong in the oceans.

Chewing through databases of brachiopod species, Finnegan surprisingly found that deepwater brachiopods—those best adapted to low-oxygen environments—were slaughtered at the first blushes of the Ordovician ice age. So too were their fellow-traveling blind trilobites of the deep.* So what was going on down there?

Today the circulation of our well-oxygenated ocean is driven in large part by the difference in temperature between the frigid poles and the balmy tropics. This engine continually brings frigid, oxygen-rich surface waters to the deep, on a worldwide conveyor belt. In a much warmer Ordovician world, circulation might have been more sluggish and the deep ocean less well oxygenated. The sudden appearance of massive glaciers on Africa, then, might have kick-started the ocean circulation, bringing a blast of oxygen to the deep.†

---

* The death of deepwater organisms in the Ordovician extinction also falsifies the gamma ray burst hypothesis.

† Not everyone agrees with this thermodynamic explanation of why Ordovician oceans were so oxygen-poor, and other reasons have been put forward, the strangest of which might have been their relative lack of fish. Fish package phosphorus in their skeletons and send it to the deep when they die. Without

For animals like the deepwater brachiopods that made the trade-off of living a boring life in a low-oxygen environment (but one with few predators), or the deepwater trilobites that might have survived the meager conditions by harvesting bacteria grown on their bodies, luck ran out when Africa froze over, ocean circulation ramped up, and these habitats vanished.

As for the mysterious graptolites—the ones who tethered to strange little sawblades and tweezers and rowed offshore like gelatinous crew teams—the changing ocean circulation might have similarly spelled doom.

"The kill mechanism really was the change in food," said State University of New York at Buffalo paleontologist Charles Mitchell.

I met Mitchell at a weeklong international symposium on the Ordovician at that most international of cities, Harrisonburg, Virginia. The city literally sits on top of the Ordovician. Caves in the area are carved out of limestone made by ancient sea life, and they're filled with graffiti left by Confederate troops hiding from the Union army. In the summer of 2015, paleontologists from around the world descended on the college town to compare notes on the period, grouse about the quality of the Virginia wines served at the conference banquets, and salute retired legends in the field like Ohio State's Stig Bergstrom (the Michael Jordan of Ordovician research).*

Mitchell was in Harrisonburg to speak to the conference attendees on his beloved and enigmatic graptolites and how they responded to the cataclysm at the end of the Ordovician (in short, not well). Deepwater tropical species—like those you can find in

---

fish, this powerful nutrient weathering out of the mountains was more readily available in the water column to fuel oxygen-hogging blooms of plankton.

* Bergstrom was honored with cupcakes decorated with icing in the shape of (I was told) "amorphognatus conodonts," and students and colleagues alike lined up to have pictures taken with the legend.

what were the sunless depths of the Ordovician ocean but is now the middle of Nevada—got wiped out altogether, but not from any drop in sea level. These animals were wiped out by the disappearance of their favorite food: the clouds of bacteria that lived in the deep, low-oxygen realm. When the oceans turned over at the start of the ice age, this low-oxygen zone was obliterated and the food disappeared, as did most of the graptolites. In the surface waters, austere and self-sufficient blooms of cyanobacteria were replaced by seaweed, which fed off of the nutrients newly welling up from the deep. In other words, the bottom of the food chain was completely replaced throughout the ocean. Seaweed might make a heartier meal than cyanobacteria, but not if you haven't evolved to eat it. Bye-bye, bug-eyed trilobites swimming through the sea.

Add these victims to the countless species left out to dry on the continents, along with the climate refugees stranded on their islands, and the mass extinction begins to come into focus. There was nowhere to hide. Not the shallows, not the deep, not the middle of the ocean.

"Extinction is really a species-level process," said Mitchell. "It's not something that happens to individuals. You and I die, but that has nothing to do with mass extinctions. A mass extinction is when everybody dies."

And everybody dies when adaptation is no longer possible.

In studying the End-Ordovician mass extinction, Mitchell was shocked by how little species were actually trying to adapt to the changing world, a potentially frightening indication that evolution isn't quite as pliable as one would hope in the face of disaster.

"We've done some work where we've actually looked at this, and the species aren't changing at all," he said. "You're in the midst of all this crisis, you would think nothing was happening. . . . Species by and large are in stasis nearly all the time, and the only time they're not is when they're in crisis. And crisis is a risky operation, right? Sometimes you make a new species, but sometimes you just go extinct."

Just as deer will never evolve to outrun hunters' bullets, mass extinctions surpass the evolutionary potential of their victims.

"It doesn't make a difference how much you want to have four eyes," said Mitchell. "If you've got no variance for eye number as a species, you're screwed."

So we now have several kill mechanisms for the end of the Ordovician, some more subtle than others. There was the draining of the seas, the cooling of the tropics, the distance between the continents, the oxygenation of the deep, and the collapse of the food chain. But we're not done killing this world yet. There was still one final coup de grâce.

Directly on top of the glacial rocks of North Africa and Saudi Arabia are black radioactive shales. If these sound ominous, they are. These are the so-called hot shales of North Africa and the Middle East—the kind that keep oil and gas multinationals up at night, eyes a-twinkling with dollar signs. The hot shales form one of the world's most important petroleum source rocks. The waters that delivered these black, oil-soaked rocks finished off what was left of the Ordovician world—seeing to grim completion what the glaciers had started. After perhaps a million years of ice, with glacial advances and retreats just like those in our own ice age, the world rocketed out of this End-Ordovician ice age into a sweltering greenhouse. The seas rose more than 100 feet, once more flooding the continents. The warm, oxygen-poor seas of the Ordovician returned with a vengeance, smothering the few surviving animals who had made the mistake of adapting to this temporary ice world. They were rewarded for their perseverance with death.

Sometimes black shales are about as close to an SOS as one

finds in the fossil record—a grim notice that oxygen is running dangerously low. They're black from being suffused with carbon from the dead sea life that sinks and settles to the bottom of the ocean, where it can't oxidize or decay. There it's left undisturbed to accumulate on a lifeless, anoxic seabed. And there it stays until it's discovered half a billion years later by a curious species of primate determined to dig it up and burn it.

Why these returning seas at the end of the Ordovician were *so* anoxic is still a matter of some debate, but one factor that might have played a role was the huge influx of freshwater spilling off rapidly melting African ice sheets. Freshwater forms a cap on top of salty seawater, and as a result, the seas become stratified, with the deeper waters starving for oxygen. At the end of the most recent ice age a few thousand years ago, oxygen in the oceans briefly plummeted from the influx of fresh glacial meltwater before slowly recovering. And today, off the south coast of Greenland, a huge mass of freshwater from the rapidly melting continent is putting a kink in ocean circulation and perhaps even slowing the Gulf Stream. Geologists like Jan Zalasiewicz at the University of Leeds think that, as humanity shoots headlong out of its current interglacial climate toward a future greenhouse of the sort not seen for tens of millions of years, we can learn a thing or two from this grisly Ordovician coda.

"The thing about the end of the Ordovician is, we're looking at a large-scale warming, sea level rise, stagnation, extinction event within what is essentially a glacial climate," he said. "So in that way it's similar to [modern] times. For the past two and a half million years, we've been looking at a system that's finely balanced between temperatures that are glacial to more or less like today's—within a degree or so. What's happening now is we're not quite at peak interglacial warmth, but we're quite close to it. But once we get higher than that we're clearly in new territory—territory that has not been seen for several million years. . . . Similarly, at the end of the Ordovician there were these large temperature and sea level and oxygenation jumps from one state to another that might be akin to the kind of dislocation we seem to be heading towards."

Nevertheless, with changes come new opportunities. Today oxygen-minimum zones in the ocean are expanding from agriculture runoff and global warming, promoting the survival of animals that have learned to hunt in this inhospitable layer—like the vicious Humboldt squid, which today is thriving in the quickly changing Pacific. But change was too rapid for life in the Ordovician to survive, as may yet happen again for life in the coming centuries.

It took 5 *million* years for the planet to fully recover from the End-Ordovician mass extinction. When it finally did, the hollowed-out ecosystem provided new opportunities for survivors to flourish. Slowly, the planet began to look a little bit more like Earth. Things with backbones—our ancestors—had thus far been unimportant players, but now they radiated in the wake of the extinction. So much for the Sea Without Fish.

Ohio State's Saltzman thinks there are lessons to be learned from the Ordovician mass extinction, even though it happened nearly a half-billion years ago on an unfamiliar planet. Of the many geochemical signals that appear in the rocks announcing the extinction, the most relevant to our day might be the wild swings in the carbon cycle, which goes haywire throughout the catastrophe. The exact meaning of these squiggles is still being debated by geologists, but the implication is clear.

"I think all you can say," Saltzman said, "is that when there are severe, rapid changes in the carbon cycle, it doesn't end well."

By the time the planet had been reborn again in the next age, life had already buried an incredible amount of carbon in the rocks that would eventually become today's fossil fuels. But biology was just getting started. Like the gun that's left on the table in the first act of a play, this one would eventually go off.

3

THE LATE DEVONIAN MASS EXTINCTION

374,359 Million Years Ago

Going up that river was like traveling back to the earliest beginnings of the world, when vegetation rioted on the earth and the big trees were kings.
—Joseph Conrad, 1899

Placoderms were the monarchs of the ocean.
—O. C. Marsh, 1877

In the past few years the United States suddenly started gushing natural gas. Scattered across the country, prospectors have drilled thousands of wells, flooding the market with cheap energy. This is the Shale Gas Revolution, and it has reorganized "the Great Game" of oil and gas geopolitics, leaving the United States less dependent on foreign energy and alone atop the world's top gas producers. The revolution was born of the technological breakthrough known as hydraulic fracturing, or "fracking," which has unlocked huge reserves of hydrocarbons and allowed drillers to suck fuel-soaked rocks dry from New York to North Dakota. The revolution has also ignited debate about the environmental and public health consequences of all this drilling, even as the country opens its spigots.

But if the effects of fracking black shale have been profound for the American economy, they have nothing on the effects that creating all this black shale had on planet Earth in the first place. For much of this newfound natural gas richness the United States can thank the horrifying mass extinctions of the Late Devonian period. More than 350 million years ago, as the seas that covered the country repeatedly suffocated, marine life died en masse, sank to the seafloor, and—much to the delight of outfits like Hess and Chesapeake Energy—eventually became natural gas.

Following the grisly end of the Ordovician (and after a brief* chapter of earth history called the Silurian), the Devonian, a major transitional period in the history of life, began roughly 420 million years ago and ended in disaster 60 million years later. In the millions of years since the "Sea Without Fish" of the Ordovician, much had changed on planet Earth. In fact, in the wake of the destruction of the End-Ordovician mass extinction, our ancestors—fish—radiated and took over the oceans. They were so successful in conquering the planet that by the Devonian period Earth had entered what's known as the "Age of Fishes." Swirling around the epic, globe-spanning reefs of the Devonian was a fully stocked ecosystem of fish predators—many terrifying and unfamiliar—and fish prey. Some of these new stewards of the ocean were even timidly testing out life on land, waddling onto shore for brief sojourns. But our piscine ancestors were in for a horrible surprise. Several horrible surprises, in fact.

---

* Only 20 million years long.

The first major deathblow of the Late Devonian mass extinctions struck 374 million years ago. All on its own, the episode qualifies as one of the top five worst mass extinctions ever, destroying 99 percent of the largest reefs the world has ever known. The reefs stretched over more than 3 million square miles, ten times the extent of modern reefs, and today their remains hold vast stores of oil in Canada and Australia. It would take more than 100 million years for reefs on the planet to recover from this massacre. But unfortunately for life on earth, the Devonian mass extinction was not a one-off disaster. The *second* major deathblow struck 359 million years ago. This final catastrophe would emphatically end the period in an icy climax, taking out the top predators on the planet: heavily armored marine juggernauts that should be finalists on any short list for the scariest animals ever.

But to understand why the planet nearly died hundreds of millions of years ago in the Devonian, we first need to explain why the United States has so much gas-rich black shale.

"I began working on the Devonian shales because they're here in Ohio," said University of Cincinnati geologist Thomas Algeo. "But then I just started wondering: why do we have so much black shale in the Devonian?"

Algeo is one of the most influential thinkers about the mass extinctions that punctuate the Late Devonian, and I met with him in his office to get a primer on this sometimes neglected period of earth history. His ideas about the Late Devonian crises have been sometimes slow to catch on in a field where many of his colleagues have been working for decades under the spell of the Alvarez Asteroid Impact Hypothesis.

"The Devonian extinction has a totally different pattern than the other mass extinction events," said Algeo. "Just in terms of duration to begin with. You're talking about something that played out over 20 to 25 million years. It's *that* long."

With at least ten extinction peaks, including the two most extreme catastrophes at 374 million and 359 million years ago—known as the Kellwasser and Hangenberg Events*—the crises that punctuate the Late Devonian are strikingly different from any of the other Big Five mass extinctions. They're especially distinct from the End-Cretaceous dinosaur extinction, which appears as almost instantaneous in the fossil record—as we might expect from, say, an asteroid impact. Nonetheless, the odd signature of the Devonian extinctions hasn't stopped paleontologists from searching for an extraterrestrial culprit.

In fact, the first deathblow of the Devonian, the Kellwasser Event, was also the first global catastrophe that paleontologists ever attempted to explain by an asteroid strike. Speaking to the Paleontological Society in 1969, Canadian geologist Digby McLaren addressed an audience of his peers, most of whom didn't even believe in the reality of mass extinctions. Since Darwin, most thought that the huge breaks in fossil life that punctuated geology were the artifact of an imperfect rock record. The idea that life could be wiped out in an eyeblink was a disreputable one—one that reeked of Old Testament destruction. The field had spent almost two centuries trying to shake off the specter of Genesis-inspired flood "geology" of the sort that now enjoys a nauseating revival in some parts of the United States. But the huge breaks in fossil life continued to haunt McLaren. He insisted that his colleagues were "trying to define out of existence" these jarring discontinuities in the history of life. In 1965 a colleague of McLaren's visited Devonian rocks in Iran and saw the same ominous break in the fossil record as he had seen in his North American rocks. Whatever had happened in the Late Devonian, it

* Also known as the Frasnian-Famennian and Devonian-Carboniferous boundaries, respectively.

was a global event. And McLaren was convinced that it required an appropriately apocalyptic cause.

"A giant meteorite falling in the middle of the Atlantic Ocean today would generate a wave 20,000 feet high," he announced to his fellow paleontologists.

"This will do," he said, endorsing the scenario for the global destruction of life visible in the rocks of the Late Devonian. The speech was received with embarrassed silence. McLaren would eventually be vindicated when researchers later found that an asteroid could play such a role in global destruction. He just had the wrong mass extinction.

When the shocking link was made in 1980 between a 6-mile-wide asteroid and the disappearance of the dinosaurs, the study of mass extinctions—long pushed aside as a disreputable fringe of paleontology—suddenly became cutting-edge. Geologists scattered to the far corners of the earth in search of extraterrestrial evidence at other extinction boundaries. McLaren's previously dismissed ideas gained new life, and the Devonian exterminations beckoned would-be asteroid hunters. Though the evidence was scant, some still claimed to have found support for catastrophic impacts near the end of the period that could have brought on such devastation.

Just as at the end of the Ordovician, some geologists think extreme cooling played a role in the crises of the late Devonian, during what was an otherwise balmy stretch of earth history. Some impact enthusiasts proposed to explain not only this apparent chill but also the unusually long duration of the Devonian extinctions and their pulsed nature by appealing to the heavens. A low-trajectory asteroid, one such theory went, could slam into Earth and kick up enough rock into orbit to form a Saturn-like ring around the planet. The shadow from the ring over the equator would create a perpetual twilight and cooling that could account for the intense extinction among tropical creatures. Then, over millions of years, the rock from the ring would slowly rain back down to earth, accounting for the duration and pulses of the extinctions. It's an evocative, if fanciful, model—and one that would have made for quite the spectacle for alien observers. But Algeo didn't buy it.

"It's improbable," he said laconically. "There are some wild ideas out there."

George McGhee Jr. of Rutgers University was one of the more enthusiastic impact proponents and published a book in 1996 outlining the case that asteroid strikes caused the Late Devonian mass extinctions. More recently, however, McGhee himself has admitted that nothing like the multiple lines of evidence at the end of the dinosaurs' reign pointing toward an asteroid impact—like an abundance of extraterrestrial dust or craters large enough to account for mass death—has yet been detected for the Devonian. Not "even," he writes, "after 30 years of searching by scientists around the world."

Algeo hypothesizes a much more down-to-earth, if equally preposterous-sounding, killer. It's one that can account, he claims, for both the deep freeze at the end of the period and all the black shale that today showers frackers in natural gas, but that in the Devonian signaled a smothering ocean unfit for animal life. Outside Algeo's office was a grooved, bulky column of stone. It was a hunk of fossilized tree—380 million years old. It wasn't there just for decoration.

"I feel pretty confident these extinctions are related to the evolution of land plants."

Kristin Wyckoff is the volunteer director of the Gilboa Museum, a modest one-room structure surrounded by farmland in the heart of the Catskills in upstate New York. The museum is filled with rusting farm equipment and sepia-toned photographs of the stately mansions of Gilboa, New York—which no longer exists. The old town of Gilboa is at the bottom of a man-made lake. Wyckoff's museum celebrates this drowned town and commemorates its tragic end, which unexpectedly brought it face-to-face with one of the most important events in the history of life on earth.

A century ago, 150 miles south of Gilboa, New York City was struggling to accommodate the crush of immigrants passing through the halls of Ellis Island and into the bewildering metropolis. The city had also just incorporated Brooklyn—long an independent enclave on the other side of the East River—into its city limits, and as a result, Gotham desperately needed more water. Naturally, it looked north, to the sparsely populated, pristine valleys of the Catskills, to build a reservoir. Unfortunately for the quaint town of Gilboa, one of the most promising places to put a dam in the entire state was literally in the center of town. The village found itself at the bottom of a valley that New York water managers projected would be one of the fastest-recharging reservoirs in the entire state, fed by over forty tributaries. Here the Scoharie River would be impounded and the beloved hamlet of Gilboa would have to be sacrificed for the greater good. The future of the town, the state decided, would be at the bottom of a lake. When New York City came to destroy the village to slake its thirst, it did so with a marked lack of bedside manner.

While collecting oral history from the former town's last surviving inhabitants, Wyckoff recalled the almost biblical exile that was imposed during dam construction. Residents came home to find ominous Xs scrawled on their doors, marking their homes for destruction.

"[One resident] said that for the last load they went back to get her doll collection in her bedroom and the house was already leveled," said Wyckoff. "They had burned it down. They didn't give you a lot of notice."

In 1926 the Scoharie River came to a halt at the gleaming new Gilboa Dam and the town was slowly submerged. One Gilboa folk historian, perhaps with only a slight tinge of melodrama, surmised that "more than one third of the population died of heartbreak. The remainder dazedly scattered to the four winds to seek abiding places until death should release them from a cruel world, which, they said, seemed to have no place for them."

Today, what remains of the old town of Gilboa lies at the bottom of the Scoharie Reservoir, whose water still pours out of Manhattan faucets. Not far from the museum, off Wyckoff Road, is a lonely graveyard filled with crooked, mossy headstones marked only by numbers: the state was forced to relocate the town's dead when the old village cemetery was submerged along with the village. More than 90 years on, the anger toward the city remains palpable. "You can still feel it," Wyckoff said.

But the construction of the dam was not without its blessings. "The one positive thing to come out of the whole Gilboa Dam is that they found these incredible fossils."

While excavating sandstone blocks to face the dam, puzzled quarriers kept jamming their steam shovels on strange grooved columns of stone. They had unwittingly discovered the first trees on the planet and the oldest forests in the history of life. About an hour west of Albany.

The state of New York had collided with the Devonian.

For almost the entire history of the earth, the continents were bleak and forbidding expanses of rock, their interiors pervaded by an empty lifelessness that would have made our home planet unrecognizable. But starting in the Ordovician, tiny plants began to establish a tenuous beachhead onshore. Thanks to its uninspiring size, this is what's known as the "Lilliputian plant world," as the pioneering sprigs of liverwort topped out at only a few inches at most. Even so, making the leap from pond scum to land plants, however modest, was an awe-inspiring feat that required major innovations among the Lilliputians to avoid drying out under a newly withering sun (like conjuring a waxy coating and tiny pores to breathe). Once plants were on land, limited real estate and competition for sunlight eventually led to an even more earth-changing innovation in plants: by the mid-Devonian, it was time to go vertical. With the development of supporting vascular tissue, trees raced to the top of the canopy, elbowing each other out of the way for sunlight.

"This is when plants go from knee-high to tree-high," said Algeo.

Where before there were no trees in the fossil record, suddenly there were 30-foot-tall palmlike jungles in Gilboa, rising over the coastal plains and wetlands of New York. These proto-trees were anchored to the earth with a tenuous hula-skirt mooring of fibers, but soon plants would aim their shoots downward as well, developing proper roots and plowing into the earth. The colonization of the land was well under way.

These first forests welcomed the world's first insects into their airy naves, and millipedes and proto-spiders skittered through tropical Gilboa. These first insects would lure fish to take their first tentative waddles onto dry land, eventually abandoning their ancestral seas altogether. After hundreds of millions of years, complex life was emerging from its crowded submarine nursery and infiltrating the bleak and lifeless continents. But it would be punished for this pioneering spirit.

When renovations to the crumbling Gilboa Dam began again in 2010, a team led by SUNY Binghamton paleobotanist William Stein was invited by the New York State Department of Environmental Protection for a peek at this world's first forest, which the state has kept off-limits to scientists and the public for almost a century—and still does today. The site was a wonderland: Stein's team found a remarkably preserved forest floor, with 200 holes where the world's first trees once stood, and they later donated several of the more than 30 unwieldly fossil tree stumps they recovered to Wyckoff's museum. The Gilboa Fossil Forest not only provides a window into a primitive ecosystem but also marks the beginning of a new planet, one whose surface would be dramatically reworked by vegetation. As Stein's team wrote in the journal *Nature* about the epochal changes kicked off by the Gilboa forest: "The origin of trees by the mid-Devonian epoch signals a major change in terrestrial ecosystems with potential long-term consequences including increased weathering, drop in atmospheric CO_2 . . . and mass extinction."

Although trees today are seen as beneficent givers of life—and plants would eventually underwrite all the flourishing of life on land to follow—these first forests on the planet might have heralded the end-times.

So how does the dawn of forests on the continents have anything to do with the alarming black shales that formed in the ocean and the extreme crises that hammered the planet in the Late Devonian?

"What happened in the Devonian is analogous to modern dead zones," Algeo says.

Today, every summer in the Gulf of Mexico, an area of the ocean up to the size of New Jersey loses its oxygen and almost everything in it dies. For its part, New Jersey suffers its own seasonal anoxia, as does Lake Erie, which saw the arrival of a toxic algae bloom so large in 2014 that it shut down the drinking supply for the city of Toledo. In 2016 the coast of Florida was battered by waves of thick, sea-life-smothering algal muck; boat owners described it as having the consistency of guacamole. The same kinds of problems afflict the greatly impoverished Chesapeake Bay,* which up until relatively recently was a biological paradise. The Chesapeake once boasted oyster reefs so extensive as to represent a navigational hazard to boats, as well as a menagerie of sea life that included "dolphins, manatees, river otters, sea turtles, alligators, giant sturgeon, sharks and rays." That roster might surprise modern pleasure boaters on the murky bay, where today one is about as likely to find a hippopotamus as a manatee. Farther afield, oxygen-poor waters beset the Baltic and East China Seas. This deadly phenomenon—runaway algae growth robbing the seas of oxygen—is called eutrophication. For the doomed animals of the Late Devonian, these tides of guacamole might have been a familiar sight.

Eutrophication is caused by too much of a good thing—an overdose of plant food. Today the Gulf of Mexico's problem starts in the heartland. When farmers in the endless rectangular patchwork of the Midwest and the Great Plains spread fertilizer rich in nitrogen and phosphorus on their crops, what isn't taken up by the plants is eventually washed into the Mississippi River. When the Mississippi empties into the ocean south of Louisiana, that accumulated Miracle-Gro spurs explosive algae growth in the open ocean. When the algae blooms die en masse, they sink and decompose, a process that uses up most of the oxygen in the water column.

Without oxygen, everything else suffocates, and the result is massive fish kills that make themselves known to Gulf Coast beachgoers in yearly biblical tides of lifeless stingrays, flounder, shrimp, eels, and fish. Underwater, dead crabs, clams, and burrowing worms litter the seabed like the casualties of an inverte-

* Chesapeake Bay, surprisingly, was formed by a giant asteroid impact 35 million years ago.

brate Battle of the Somme. It's easy to see how scaled-up, global, and persistent eutrophication episodes could be apocalyptic, as they might have been in the Devonian.

Today these algae blooms* and dead zones around the world are spreading, driven mainly by development and the growth of industrial agriculture.† But with little evidence for a Devonian Monsanto, another explanation is needed for an ancient influx of plant food to the seas. The explanation offered by Algeo and his colleagues is the plants themselves: they dug into the earth with their roots for the first time, they say, broke down the rocks, and liberated nutrients—like phosphorus—which then washed out into rivers and poisoned the oceans with algae and phytoplankton food. The resulting nutrient deluge spurred huge plankton blooms that robbed the seas of oxygen and, ultimately, created all that black shale.

The trees in Gilboa were really just weird, enormous weeds. Though they had thick trunks like palm trees, they were more closely related to horsetails and ferns. The first trees we might recognize as such appeared only later in the Devonian, when *Archaeopteris*‡ arrived on the scene.§ Resembling a slender cedar tree, it towered more than 100 feet in the air. To support this impressive stature, *Archaeopteris* had the world's first deep-root system. It secreted organic acids and dug into the virgin earth—physically and chemically attacking the continental rock, and breaking it down as it spread.

These trees formed the first soils—soils that then washed into

---

* Besides using up oxygen, toxic algae blooms live up to their name. An episode of psychotic seabirds that ingested algal neurotoxins in Monterey Bay served as the inspiration for Alfred Hitchcock's *The Birds*. Recent research on human populations surrounding New England ponds has even linked toxic algae blooms to ALS (amyotrophic lateral sclerosis) hot spots.

† And exacerbated by global warming.

‡ Not to be confused with *Archaeopteryx*, a famous transitional birdlike dinosaur with feathers. Paleontologists don't make it easy for the casual follower.

§ This tree was still quite strange, using spores to reproduce rather than seeds.

rivers and streams and ultimately out into the shallow seas, flooding the ocean with prehistoric Miracle-Gro. As the kingdom of the *Archaeopteris* quickly spread across the planet, downstream the nutrients they released from the rocks spurred blooms in the ocean, which suffocated life in the seas, much as industrial fertilizer does today. These plankton blooms are evident in the rocks as huge carbon burial events, a signature of primary productivity in the seas run wild. The carbon from these deadly tides is the same being fracked today.

Compounding the problems in the Devonian were the restricted openings to the open ocean from the strange seas that covered the continents, which made it even more difficult to flush out this torrent of nutrients coming from the land.

"Today these dead zones in the Gulf of Mexico are taking place on an open continental shelf—there's no restriction there to speak of. So it's possible even in those conditions too," Algeo said. "But in the Devonian, there really were prime conditions for developing anoxia."

This was the ultimate fish kill.

Further tree innovations—like seeds at the very end of the Devonian, just before the second wave of mass extinction—allowed plants to push even farther into dry environments and survive. What had begun in the marshy margins around streams and ponds in places like Gilboa was a full-scale invasion of the earth by the end of the Devonian. Algeo thinks that these sorts of biological innovations—trees, roots, seeds, and so on—account for the pulsed nature of the Late Devonian extinctions.

"The spread of vegetation wasn't necessarily a uniform process occurring over 25 to 30 million years," he said. "It may have occurred in pulses. For instance, one group of plants would evolve the ability to spread farther inland and adapt to harsher conditions. That happened very quickly. Then there would have been a period of stasis, [and] then another paleobotanical development would have triggered the next crisis."

If the early forests had *only* flooded the Late Devonian oceans with excess nutrients and sapped the seas of oxygen, the extinctions might not have been so ruinous. But trees have another trick. They suck up enormous amounts of carbon dioxide. This is one reason there's so much hand-wringing today over the destruction of the Amazon rainforest as the planet fumbles with its carbon budget in a warming world. One (contentious) theory for the recent cold snap that lasted from roughly AD 1500 to 1800 invokes the mass death of American Indians and the reforestation of North America after centuries of slash-and-burn indigenous agriculture. As growing trees reclaimed the post-Columbus continent they might have kicked off a brief chill by hoarding vast quantities of atmospheric CO_2. After all, trees don't grow from the ground up, but from the air around them.

In the Devonian the original forestation of the barren continents took place on an altogether different scale. As trees spread out over the world, carbon dioxide in the atmosphere would eventually drop by more than 90 percent. Moreover, the carbon locked up in the world's first forests and soils was compounded by the huge amounts of carbon then being buried in the anoxic seas by nutrient-fueled plankton blooms. Unsurprisingly, all this carbon burial had major consequences for the climate, and for life. It got very cold.

"Each of the two largest mass extinction events in the Late Devonian were associated with sharp cooling and continental glaciation," said Algeo.

If he's right, not only was the planet punished with suffocating seas, but it was throttled by icy climate paroxysms as CO_2 plummeted once more.

Unfortunately for Algeo, there's some conflicting evidence: the first deathblow of the Late Devonian, the Kellwasser Event, which brought low the world's reefs, is still a rather cryptic episode. One reason for this ambiguity is that this first wave of extinction took place over a million years and had as many as five deadly pulses within it. A lot can happen in a million years. Human beings are only 200,000 years old.

As such, the causes of this first wave of mass death remain, as the academic papers put it gingerly, "highly controversial," and different papers on this earlier extinction can read like descriptions of altogether different events. Nevertheless, Algeo's case for brief ice ages during the Kellwasser Event, caused by the spread of plants and plummeting CO_2, though far from ironclad, is grounded in some evidence: oxygen isotopes from the teeth of tiny eel-like animals point to brief, but wildly precipitous, drops in tropical sea temperatures of 5 to 7 degrees Celsius.* Elsewhere, eroded rocks as far afield as China and western Canada point to a dramatic fall in sea level in the Late Devonian, while cold-adapted creatures seem to have preferentially survived in the extinction's aftermath and closed in on the tropics.

But the evidence is contradictory, and one specter that lingers uneasily over the whole period is what looks like the eruption of a major volcanic province in Russia that would have been capable of inflicting all sorts of chaos, including extreme global *warming*. Indeed, evidence exists for this warming, as well as for huge surges of sea level *rise* during the extinction pulses. The reason for all this confusion is the problem of precisely dating rocks, the fragmentary nature of the fossil record, and ultimately, the very human job of interpreting data.

---

* In case you were wondering how that works, animals incorporate different ratios of oxygen isotopes into their skeletons at different ocean temperatures.

David Bond at the University of Hull in England doesn't doubt that the transformation of the planet by plants could have been traumatic for life on earth.

"I quite like Tom [Algeo]'s hypothesis," Bond said. "Each of these steps in the development of the bigger plants, they do seem to happen at the same time as these global anoxic events, so it's quite interesting. It's quite a neat idea."

But, as for a glaciation at the first major extinction event of the Late Devonian, Bond doesn't see it. He thinks that evidence for cooling and sea level fall in the rocks likely postdates the extinction and therefore couldn't have caused it. As for the temperature data indicating sharp cooling, Bond thinks that it's unreliable—tampered with by eons in the earth.

"I think it's true that there probably is sea level fall and glaciation, but the key is the timing," he says. "If it happens after the mass extinction, it can't be the cause."

Instead, Bond and others propose massive sea level *rise* and global *warming*—perhaps from volcanic carbon dioxide issuing from Russia and Ukraine—which brought the anoxic seas of the Devonian surging onto the shelves, wiping out almost everything in the ocean.

The eruptions have yet to be precisely dated to the mass extinction, but they are close enough to raise eyebrows. And they were eruptions of the sort that, as we'll see later, played an outsize role in every mass extinction to follow.

So at this point we don't know what caused the Kellwasser Event, one of the most important and devastating episodes in the history of life on earth. Perhaps, rather than a straightforward chill or punishing furnace in this first mass extinction of the Devonian, it was instead rapid, wild swings in climate, between fire and ice, that doomed life. For students debating whether to get a degree in geology or paleontology, all this uncertainty about such a pivotal moment in earth history should offer some encouragement: there are still big questions to answer.

It looks as though Algeo is right, though, about the second major deathblow of the Devonian, the Hangenberg Event—a disaster that brought the period to a devastating close and wiped out a terrifying suite of monstrous fish. It seems almost certain that the planet was briefly, and catastrophically, choked with ice.

"Seed plants come in and spread very rapidly at the end of the Devonian," said Algeo, "and that seems to be a trigger for the cooling and glaciation that take place at the end."

The world bloomed and the planet froze. And the evidence for this icy global disaster that terminated the Devonian is easy to find. Drive through western Maryland and you need only look out your car window.

In 1985, in his first year with the Maryland Geological Survey, state geologist David Brezinski drove out to the Alleghenies in western Maryland to inspect a massive road cut that highway workers had just blown through a ridge called Sideling Hill. The artificial cliffs revealed a layer cake of rocks, laid down in shallow seas and swamps. The layers were later contorted by continental collisions into a huge U that, if traced outward today, points toward invisible mountains that have long since been eroded away. The artificial canyon on Interstate 68 has become something of a tourist attraction for rubbernecking motorists, but for Brezinski—a scientist experienced in studying such exposures—the cliffs were simply spectacular.

The road cut was layered with the sorts of sandstones and coals that one might expect to be laid down in a warm ancient coastal environment, but at its base the rocks were suddenly interrupted by something completely out of place.

"It's basically a mudstone, but with huge boulders in it—sometimes a meter or more across. I had never seen anything like it," Brezinski told me. "I couldn't understand it."

The rocks looked like they were left by glaciers, but that couldn't have been right. This whole section of rock was long thought to have formed underwater in a tropical world. Geologists rationalized that this odd jumbled rock that interrupted it must have been from some sort of localized submarine landslide.

"But what I found out in western Maryland and western Pennsylvania," said Brezinski, "is that everywhere you went and looked closely at that same level, you found this rock. It wasn't localized at all."

Brezinski knew this wasn't the work of an underwater rockfall. And it wasn't the debris from an asteroid-induced tsunami, as others suggested. This was the work of glaciers on dry land.

"For a long time people kind of just ignored this evidence because they knew the Devonian to be a warm world," he said. "And this didn't fit that."

As strange as it was, the evidence kept adding up. In 2002, Devonian pebbles were discovered bearing the telltale streak marks of having been ground up by the churning march of glaciers. In 2008, a 3-ton granite boulder was found lodged in the Devonian shale of Kentucky—the only plausible explanation for which was having been dropped there by an iceberg. Underneath Cleveland suburbs and in German riverbanks were huge river canyons, now buried in sand and shale, that once carved into the dry Devonian seafloor when the ice advanced and the seas fell.*

Brezinski found that the strange jumbled rock that he first encountered in the Maryland I-68 road cut extended for 250 miles,

* Similar bygone river canyons, like the underwater Baltimore Canyon off the coast of Maryland, are remnants of the most recent ice age, when sea levels dropped 400 feet and rivers cut through the dry continental shelf.

from northeastern Pennsylvania through Maryland and into West Virginia. Glacial rocks were already known from locations then closer to the poles, like prehistoric Bolivia and Brazil, but the Appalachians were almost in the tropics in the Devonian. It all pointed to a geologically brief, but catastrophic, glaciation. The Devonian world, it seems, was finished off by hypothermia.

The work done by Brezinski, whose office is in a Baltimore neighborhood that had seen three murders that year, and his colleagues in western Maryland and Pennsylvania "really changed a lot of thinking on the Late Devonian," he said. "I think it's largely swung in the past five years. Even Tom Algeo wasn't a big believer in the evidence for glaciers around here, but we took him on a trip in 2006 to Pennsylvania and he said, 'All right, well, this is pretty compelling.'"

Brezinski brought me and a group of geologists and paleontologists out on a field trip to the famous road cut at Sideling Hill for a little mass extinction tourism.

As should be clear by now, perhaps no national project has had more of an incidentally beneficial effect on science than the Interstate Highway System has had on geology. In the eastern half of the United States, where malls, houses, and photosynthetic glop obscure the continent's amazing fossil wealth, highway road cuts often provide lone windows into the deep past. But as one geologist told me, if Dwight Eisenhower—as the highway system's father—was a great friend to geology, then Lady Bird Johnson—whose pet Highway Beautification Act involved covering many roadsides with greenery—was persona non grata.

"Did anybody bring any acid?" a geologist asked the group as we marched down the shoulder of I-68.

"I did!" responded another. I would learn that this surprising question was customary among geologists. The acid is used, somewhat disappointingly, to take layers of grime off the rocks.

One of the paleontologists who tagged along on the field trip was the University of Pennsylvania's Lauren Sallan, one of the world's top paleontologists of ancient fish—if not *the* top. For her, the trip to see these glacial rocks on the side of the highway firsthand was something of a pilgrimage.

Sallan has struggled to impress upon her colleagues the severity of the extinction that capped the Devonian. Where other mass extinctions are known mostly from their effect on invertebrates like brachiopods or even plankton, the mass extinction at the very end of the Devonian was an outsize slaughter of big, charismatic vertebrates. It wiped out an astonishing 96 percent of this group that in the Devonian lived in the water but today includes almost all of what we tend to think of as wildlife: dogs, whales, lizards, snakes, sharks, elephants, frogs, ourselves—you name it.

"The end of the Devonian is the worst extinction ever for vertebrates," she said.

It was even worse, she said, than the End-Cretaceous mass extinction that took out the dinosaurs and most other life on earth.

"At least at the end of the Cretaceous the fish don't all die."

Brezinski brought us to the odd glacial interval at the base of the road cut that marked the catastrophic end of the Devonian and invited us to take souvenirs. If the severity of the End-Devonian extinction is only slowly seeping into academic consciousness, the idea that a disastrous glaciation happened here as well has similarly failed to fully penetrate the field. As is often the case, the specialization of the university has kept colleagues blind to the research taking place the next office over. Standing on the side of the highway in western Maryland, Sallan held a piece of glacial rock in her hand, stared into the middle distance, and announced:

"My life is now complete. I have a piece of the extinction. It's a piece of the glaciation. I can show it to people in my office and say, 'Look at that. It happened.' I keep telling people it was a devastating mass extinction because there were glaciers at sea level in the tropics, and most people don't believe me. So now I guess I can just throw this rock at them, and then when they're in the hospital they'll believe me."

Sallan's prickliness about the underappreciated event stems, in part, from her love of placoderms: an eclectic assortment of heavily armored fish that ruled the Devonian. If you've never heard of placoderms, most likely that's because they aren't around anymore—but, Sallan says, that's not their fault. Although they might not occupy the same real estate in pop culture, placoderms were to the Devonian what dinosaurs were to the Jurassic and Cretaceous.

"Placoderms were everywhere. They did everything. Then they're just gone," she said.

To understand why Sallan sounded almost heartbroken over the disappearance of these peculiar creatures, I booked a ticket to Cleveland to meet them. There, literally falling out of the riverbanks, are the remains of some of the most vicious and fearsome predators in the history of the planet.

Where Cincinnati's recovery from Rust Belt decay is evident in a redeveloped riverfront and a hip corridor of restaurants, breweries, and bars, Cleveland's charms—a downtown bisected by an industry-lined (and once flammable) river, strewn with parking lots, and crowned with a formerly regal department store turned retirement-fund-sucking casino—are harder to appreciate. To a paleontologist, however, the city's a dream. It's built on a foundation of sea monster bones.

"We tend to anthropomorphize these things, but come on, it looks evil," said Cleveland Museum of Natural History curator Michael Ryan. I met Ryan in the bowels of his museum's collections, where the giant ossified heads of the killer placoderm *Dunkleosteus* surrounded us in various states of reconstruction. The 360-million-year-old ocean-prowling juggernaut was named for Ryan's predecessor at the Museum, David Dunkle, and the giant skulls were pulled from local Cleveland riverbanks, which have produced staggering specimens for museums around the world. *Dunkleosteus* is invariably displayed with jaws agape and a frozen expression of blank malice. Dinosaurs, *The Economist* once playfully wrote, were "reassuringly extinct." The description, if it fits any creature, fits *Dunkleosteus*.

Encased in a sturdy helmet of bone, *Dunkleosteus* was the unquestioned lord of the Devonian seas, and one of the world's first apex predators with a backbone. The length of a Winnebago, it dwarfed the meek, yard-long sharks that had recently evolved and wisely assumed a peripheral role in the ocean. Unlike the teeth of today's more familiar fish, the sinister *Dunkleosteus* had a pronged guillotine of self-sharpening bony plates protruding from its head-mounted armor, unlike anything alive today. It used these blades to shear flesh, break bones, and spread dread through the water column like an oil spill. Unlike animals that move only their lower jaws to feed, *Dunkleosteus* had a huge muscular hinge on top of its head that allowed it to swing its bladed mouth open like a snapping turtle undergoing an exorcism. The suction created by this movement was so intense that *Dunkleosteus* would actually inhale animals into its gaping mouth before slamming it closed with the most powerful bite ever in the history of fish. Its redoubtable bony armor plating—more than an inch thick—makes sense only in a world with other *Dunkleosteus*. Unsurprisingly, one *Dunkleosteus* head shield found in the Cleveland shale bears gouges that, according to a paper by Ryan and his colleagues, are "best interpreted as bite marks inflicted by . . . another *Dunkleosteus*." Less fortified animals stood no chance, and *Dunkleosteus* showed no mercy. It's even been suggested that mangled fish parts found surrounding *Dunkleosteus* in the fossil record are indicative of "gluttony" on the part of the overindulgent sea monster, the regurgitated leftovers of a murderous bacchanal.

"What else is there to say about them?" Ryan said, gesturing toward one of the glowering, armor-plated heads. "I mean, they look evil, right?"

They do.

It may be less than a coincidence that at the same time animals like this were terrorizing the seas, our paddle-limbed fishy ancestors were making their first tentative forays onto dry land. As American Museum of Natural History paleontologist John Maisey told *Nature,* our ancestors "did not so much conquer the land, as escape from the water."

That is, they were literally scared onto terra firma.

University of Chicago evolutionary biologist Neil Shubin writes, "The strategies to succeed in [the Devonian] were pretty obvious: get big, get armor, or get out of the water. It looks as if our distant ancestors avoided the fight."

"These guys are designed to do one job, which is to eat other things," Ryan said, studying one of the massive Cleveland skulls in his office. "They probably did it very well."

Set against the back wall of Ryan's collections are rows of shelves containing enough shark and *Dunkleosteus* fossils recovered during the construction of Interstate 71 in Cleveland to keep him and his colleagues at the museum busy for several lifetimes. As highway workers were paving the asphalt rivers of the Midwest for the endless migrations of tail-finned cars then being spawned in Rust Belt furnaces and assembly lines, the Cleveland Museum of Natural History continuously pulled *Dunkleosteus* head shields and ancient sharks from the rubble in the highway's wake.

"We had volunteers and crews that, essentially every week, would go out to look at the spoil piles that they were digging up and just pull out anything that had bones in it," he said. "They just dumped it to the side. Before the trucks would take it away, we'd have crews going out there literally just pulling out all the stuff they wanted."

While *Dunkleosteus* enjoyed its uncontested dominion over the seas of Cleveland, its armored brother and sister placoderms took on supporting roles—and a wild variety of forms. Some are dead ringers for stingrays; others look more like fighter jets with fish tails. But the most abundant fossil fish from this era is another armored placoderm called *Bothriolepis*. Again tiled in bony plates, it's a bizarre creature that resembles a headless turtle with jointed switchblades projecting from its sides. These are its unlikely fins, which Sallan compared to crab claws.

"It's almost insectlike," she said. "It's really strange."

Along with the even more bizarre armored jawless fish that were swimming alongside—some looking rather like baroque chainsaws—the Devonian tides teemed with bony boomerang heads, extravagant spikes, and unrecognizable creatures with armored, winglike projections. Although it was the age of fishes, if an ocean dominated by bony Frisbees like *Bothriolepis* or psychotic torpedo Cuisinarts like *Dunkleosteus* sounds unusual, it should.

"Placoderms are the only group of vertebrates that have gone totally extinct," said Ryan. Where birds carry on the legacy of the dinosaurs, Ryan said, the placoderms have no descendants. "There are no living relatives."*

But killing *Dunkleosteus* and friends wasn't easy. "Placs" (as paleontologists lovingly call placoderms) weathered the pulses

---

* Lauren Sallan disputed this claim: "They don't have a common ancestor independent of other jawed vertebrates. One group of placoderms became modern jawed vertebrates, so it's the same relationship between dinosaurs and birds."

of ocean anoxia that continually beset the Late Devonian, and they even limped on past the first wave of death that claimed the world's entire reef system, enduring the loss of half their kind in the Kellwasser Event. But after the final blow that ended the Devonian once and for all—that pulse of anoxic oceans and the brutal ice age perhaps brought on by forests—there were no more of the terrible armored fish on Earth ever again.

"There were seventy species of *Bothriolepis,* it was everywhere on Earth, it was 90 percent of fossil abundance all the way until the end, and then it was just dead," said Sallan.

It was once thought that placoderms went extinct because they were gradually outcompeted by the newly evolved sharks and more modern "ray-finned" fish (think of almost every fish you know). Time-lapse animations of life on Earth typically reflect this bias, shoehorning placoderms somewhere between simple fish and landlubbing reptiles, and suggesting that they were primitive and ultimately unsuccessful experiments early in the pageant of evolution—obsolete stepping-stones to modern biodiversity. The idea visibly offended Sallan when it was brought up. Along with the near-total extinction of the so-called lobe-finned fish (fish like coelacanths, which today are zoological curiosities but in the Devonian shared top billing with placoderms), the end of the Devonian wasn't an orderly transfer of power in the oceans, but a violent overthrow wrought by mass extinction.

"The placoderms and lobe-finned fish all survived until the very end," said Sallan, as if describing the heroic last charge of a general killed in battle. "Placoderms were totally dominant. They filled almost every niche in both freshwater and marine realms. Yes, right alongside them were the early members of the ray-finned fishes and the early sharks, but the placoderms completely outnumbered them. So the whole idea of placoderms being primitive is a bias caused by the fact that they're extinct. The reason things like placoderms and lobe-finned fishes seem primitive is because they got wiped out by this extinction. There's no reason whatsoever that we couldn't have a primarily placoderm, lobe-finned fish ecosystem now, and all the years leading up to now. Sharks and ray-finned fishes might very well be extinct by now because they were minor to begin with. In an alternative scenario, maybe they would remain minor characters and get completely wiped out at [the next extinction] and we'd still have placoderms today."

Though the story was ending for many creatures at the end of the Devonian, other stories were just beginning in this brave new world. I drove into the wilds of central Pennsylvania to see where one of the greatest transitions in the history of life happened at the very end of the period. Here in the middle of the Keystone State, 360 million years ago, was an Appalachian Amazonia where rivers and oxbow lakes cut through forests of *Archaeopteris* as water streaming off the towering Catskills eventually spilled out into the sea somewhere near Pittsburgh (beyond which was the forbidding domain of *Dunkleosteus*). After 200 million years in their nurturing ancestral seas, fish were emerging onto dry land and beginning to adapt to life on the banks of these quiet river bends and lakes. This is the story of our ancestors: if you traced your family tree back far enough, eventually you might meet one of these brave fish.

These first terrestrial footsteps were truly heroic achievements, similar in many ways to our own species' tentative, and equally hazardous, first leap into the heavens in the past century. Just as outer space today beckons a booming humanity increasingly finding its tiny but comfortable world cramped and unable to support its ambitions, the vacant but unforgiving ecospace on land in the Devonian called forth a band of bold explorers escaping a crowded sea. Fatal dehydration, the crushing effects of gravity, searing radiation from above, and the gasping transition to thin, unsatisfying air were but a few of the nontrivial challenges faced by these brave Devonian pioneers. The sea was a comfortable nursery, one that our ancestors left with all the trepidation of an adolescent who has stayed too long in his childhood home.

Hynerpeton, named for the town of Hyner, Pennsylvania, is one such ambiguous fish in transition. By the end of the Devonian, it had lost its gills, breathed only air, and had developed large shoulder blades that would have connected to muscular arms. Although its name means "crawling animal from Hyner," this amphibious creature probably spent most of its life in the water. But calling it a fish doesn't seem quite right either. *Hynerpeton* was what's known as a tetrapod. As are you and I.

I followed Google's directions for the Red Hill Field Lab and Fossil Display near Hyner, where *Hynerpeton* was discovered in the 1990s by University of Chicago evolutionary biologist (and *Your Inner Fish* author) Neil Shubin. When it was discovered, it was the oldest such tetrapod ever found in North America. I figured such a momentous find in earth history would demand an appropriately momentous commemoration in this rural town. But when my phone told me I had arrived, I found myself instead at a town hall that was closed for the weekend. Confused, I asked a bystander if there was a fossil museum somewhere in town.

"Yeah, Doug keeps the fossils up there," he said, gesturing to the town hall. "He's probably home, since it's Sunday, but I'm sure he'd just be tickled to show them to you."

I was told to drive down the street to a gas station, where there would be a phone book behind the cashier and Doug's number in it. I did as I was told. But after finding the museum nonexistent and myself holding a phone book and listening to a busy signal in a gas station, I became discouraged and started driving out of town. Just then, a huge red cliff came up on my left. Halfway up was an older man hacking away at the cliff face with a rock hammer. I pulled over to the side of the highway.

"Are you Doug?" I asked, leaning out the window.

"Yep," he said.

I explained that I was writing a book about mass extinctions, and he pointed up at the cliff.

"We find *Archaeopteris* here," he said. "You're probably hearing the rumor that that was one of the causes of this extinction."

"Yep," I said.

Doug Rowe is a retired mechanical engineer and amateur paleontologist who rents the top floor of the local town hall for $1 a year. There he stores a collection of fossils worthy of the world's top natural history museums. He makes visitors sign a guest book, and the roster of signatures in his makeshift museum-cum-field-station reads like a roll call of the world's top grad students and paleontologists. Every year they make pilgrimages to this tiny Pennsylvania town, where some of the earliest fish came onto land and stopped acting like fish. Standing at the base of the roadside cliffs, Doug pointed at the red, dusty rocks and, with his finger, traced old river channels and standing pools of water, visible only to geologists. The rocks were filled with fish and plant debris. Here we found parts of a shark skull along with fish scales and the fangs of a monster lobe-finned fish, called *Hyneria,* that Doug estimated was 12 feet long.

In the Devonian, some of the lobe-finned fish became tetrapods like *Hynerpeton,* and their lobed fins became the arms and legs (and wings) of every land-dwelling vertebrate on the planet thereafter (in a modestly successful spin-off project). But the ones who stayed in the water were decimated. Today there are very few lobe-finned fish left on the planet. Though they would linger on past the Devonian, they had withstood a blow from which they would never recover. As George McGhee Jr. writes with humorous understatement about the fallen group, "Today [they] are represented only by three genera of living lungfishes, one genus of coelacanth, and, of course, ourselves." And coelacanths just barely make this list, having been thought to be extinct for tens of millions of years before one was caught off the coast of South Africa in 1938. It was one of the most shocking finds in the history of biology. As American Museum of Natural History curator Melanie L. J. Stiassny recounted, "It would be like someone calls up with a picture of a *Tyrannosaurus* saying, 'This was running around the vegetable patch. Is it interesting?' Yeah, it was."

Modern lobe-finned coelacanths even have vestigial lungs and carry around portions of DNA that can stimulate the growth of limbs. In fact, coelacanths are more closely related to you and me than they are to other fish. But if coelacanths had decided to stay in the water in the Devonian, it might have been the wise thing to do, for the pluck of these terrestrial pioneers that pushed onto land was rewarded with ruin.

"The latest Devonian was the exact wrong time to try to go onto land," said Sallan. "They almost get totally wiped out at the extinction."

Tetrapods all but disappeared for 15 million years after the extinction at the end of the Devonian. Before the disaster, some tetrapods had eight fingers, others six, others five—and they pursued all sorts of different lifestyles. There were freshwater tetrapods and tetrapods that swam in the sea. But after the gauntlet of ice and anoxia that capped the era, only freshwater tetrapods survived, and stranger still, only those with five fingers. As McGhee points out, that you aren't holding this book with fourteen fingers may very well be an artifact of the evolutionary bottleneck at the end of the Devonian.*

---

* Apologies to polydactyl Amish and industrial machinists who may exhibit digit variability unrelated to the Devonian-Carboniferous bottleneck.

The placoderms would be entirely eliminated by the convulsions at the end of the Devonian, but our audacious ancestors barely fared any better. Calling something a "success story" in the aftermath of the indiscriminate slaughter of a mass extinction merely designates the lucky few who only *almost* died.

There are a few ways to render a planet all but lifeless. One way is to kill everything—throw an enormous asteroid at a planet, or an ice age, or a period of extreme global warming, and so on. But there's another part of the equation, the flip side of extinction: speciation. If extinction rates go up but so too does the number of new species evolving, then the new species step in to fill the gap, and it's basically a wash. The strange thing about the Late Devonian period is that animal life seemed to be sapped of this creative resilience: the rate of new species being generated dramatically fell off as the drumbeat of extinctions carried on. Ohio University paleontologist Alicia Stigall has dubbed this event a "mass depletion" rather than a mass extinction. And the key to this Devonian mass depletion was foreign invaders.

On top of everything else that was going wrong in the Late Devonian, the ancient oceans were beginning to close and landmasses that had long been separated were drawing closer together—and would eventually form the supercontinent Pangaea. As these landmasses approached and sea levels pulsed up and down, weedy species spilled into new environments where they weren't welcome. A diverse world of quirky locals was slowly becoming a world of stultifying global sameness as invasive species spread and suppressed the generation of unique regional fauna. Both invertebrates—like Stigall's favorite, the brachiopods—and vertebrates like fish became more homogeneous in the Late Devonian. Add this to the climate and ocean stresses then punishing the planet and the loss of most life on earth begins to seem all but inevitable.

"Extinction is about killing stuff, and it's really easy to kill stuff," said Stigall. "I mean, honestly, you just screw with your environment and everything that lives right there dies. So it's pretty easy to do that, and there are lots of kill mechanisms. But stopping speciation is different. So yeah, I think Tom Algeo's idea about the evolution of land plants is a great kill mechanism, but it doesn't necessarily explain why we don't have new species forming. Building biodiversity is really a different process than destroying biodiversity."

Stigall didn't hesitate to compare the homogenization of the Devonian environment to the modern day, when invasive species are being transported by humans all over the world, creating a sort of artificial biological Pangaea. Mainland rats dominate the ecosystems of remote Pacific islands, and Russian zebra mussels can become such a pest in the Great Lakes that they clog the pipes of municipal water treatment plants. Where there were once distinctive local plant ecosystems, there are now continent-wide monocultures of corn and soy. In her paper describing the Devonian "mass depletion," Stigall concludes:

"The modern combination of habitat destruction coupled with species introductions is, therefore, likely to result in total biodiversity loss that may be even greater than that experienced during the [End-Permian mass extinction]." The full terrifying weight of this statement will become apparent after the next chapter.

So it seems as though there might have been many authors of the Late Devonian crises. Earth system cycles were jackknifed by the spread of trees, glaciations, volcanoes, eutrophication and ocean anoxia, invasive species, and more. As far as kill mechanisms go, it isn't all that elegant. But perhaps this is to be expected.

"Since mass extinctions have only happened a few times in the history of the earth, it could just be like the worst thing of all time," said Lauren Sallan. "Everything lines up, everything comes up snake eyes, and you get a mass extinction."

But if researchers are still teasing out the many authors of the Late Devonian destruction, one obstacle to progress in our knowledge of this transitional period of life on earth has been, amazingly, a lack of interest.

"The Devonian research community is a little anemic, frankly," said Thomas Algeo. "It's hard to get enough people together for meetings. We tried to do a special issue on the Devonian and didn't get enough papers to make it fly. There aren't enough people actively working on it."

Depressingly, while research on the pivotal Devonian period languishes, only a few miles from Algeo's office is the Creation Museum, a bizarre evangelical funhouse where glassy-eyed schoolchildren are told that the Earth isn't much older than the pyramids and shown dioramas of tyrannosaurs boarding Noah's Ark. Awash in donor money—and even state tax breaks—the Creation Museum is expanding.

Returning to Cleveland some 359 million years ago, the tepid seas are gone, dry land is underfoot, and huge river canyons cut through the frigid wastes. A few miles further south, rafts of ice float just offshore in the shrunken anoxic seas. The distant Appalachian titans now spill out glaciers from their rugged gates. And still further south, on the southern supercontinent, are howling expanses of white. Underneath the glacial stillness are the remains of the placoderms, every last one of them dead after a 70-million-year stewardship of the earth's oceans. Also gone are the great forests of *Archaeopteris*, perhaps doomed by their own success. The greatest reefs the world has ever known are long since dead, broken, and buried. Close inspection of the seawater reveals that even the plankton is a shade of its former miniature splendor, while squidlike animals that bobbed under the surf in coiled shells have all but disappeared. On land a harsh wind blows through the surviving shrubs—the architects of all this devastation, as well as the authors of all the flourishing of life on land to come. With atmospheric carbon dioxide bottoming out, we are at the opening salvo of the longest ice age in the history of animal life: the 100-million-year Late Paleozoic Ice Age.

In the aftermath of the Devonian, facing an impoverished ecosystem and the total extermination of predators like placoderms, the ocean floor exploded in gardens of plated sea lilies distantly related to starfish (called crinoids), like flowers rising from a cemetery. The Devonian is known as the age of fishes, but its miserable aftermath is referred to by echinoderm enthusiasts—less catchily—as "the age of crinoids."

If, as Algeo claims, plants were underwriting the deadly disruptions that pulsed throughout the Late Devonian, he thinks there are lessons to be learned.

"If land plants are driving the Late Devonian biocrisis, that represents not just a completely different extinction mechanism than all the others, but one that's related to evolution itself. It means that you can actually have evolution create such dynamic changes that it results in a crisis for the rest of the biosphere. I think, in that sense, it's closest to what's going on today with the impact that humans are having."

Like the first trees, we are extraordinary in the history of life for our ability to radically alter the geochemical cycles of the planet, with dramatic consequences for the climate, the oxygenation of the oceans, and life on land and in the sea. And there's something more than a little poetic in our doing so by digging up and igniting the carbon-rich life buried in black shale by the Late Devonian mass extinctions.

"What's happened is not that humans have evolved right now, since hominid evolution has been going on for 6 or 7 million years," Algeo pointed out, "but rather that we've evolved to the point where our technology is causing havoc on the planet's surface. It's an analogous development."

When I walked out of Algeo's office and into a swelteringly hot April Cincinnati day, I considered the *Eospermatopteris* tree stump outside his office and what he had told me.

"We are the trees of the Devonian," he said.

"Is this where the wharf used to be?"

The old woman grabbed my arm as I walked back to my car from the Joggins Fossil Cliffs in Nova Scotia.

"That's what they said on the tour," I offered.

"And was that where the McCarrons River dam was?" She pointed to an imaginary spot farther down the beach.

"Sorry, I'm not sure."

The old woman sighed. As I turned to walk up the stairs she began again.

"Many moons ago this is where I learned how to swim." She shook her head. For some time after, she stayed hunched over the railing, staring into the distance, straining to reimagine her lost world. While her youthful haunts of only a few decades past had been erased by the neglect of the tides, a 315-million-year-old tree trunk stood upright in the coal-packed beach cliffs. Broken slabs from these cliffs revealed what looked like tractor-tire tracks: the fossil footsteps of preposterous 8-foot-long millipedes. Other rocks in the cliffs contained dragonflies the size of seagulls, and in the hollows of some of the fossil tree trunks were the remains of the first reptiles to spend their entire lives on land, finally divorced from their ancestral seas. Though the Devonian was over, these rocks recorded a new world reshaped by the innovations of that age, transformed by plants, and conquered at last by animals.

"How things change." The woman sighed.

The age that followed the Devonian was a more familiar world. Tetrapods now laid eggs with shells, which allowed these former fish to finally bring their reproduction fully out of the water and lead their entire lives on land. Eardrums allowed them to hear. Trees, meanwhile, were still burying carbon like mad. The period after the Devonian is known as the Carboniferous, and it supplied most of the world's coal. Burning coal, of course, makes the planet warmer by releasing carbon dioxide, but burying it in coal swamps hundreds of millions of years ago cooled this ancient planet even more. Though it was jungly here in tropical Nova Scotia, glaciers were now a lasting feature at higher latitudes. The flip side of a cool, low–carbon dioxide world is one that is positively suffused with oxygen, exhaled by the newly established plant world. Although it might not sound like much, as trees were buried in the coal swamps of the Carboniferous, oxygen spiked to as much as 35 percent of the atmosphere (compared with 21 percent today). This oxygen-rich environment explains the tractor-tire millipede tracks and the gull-sized dragonflies in the Nova Scotia rocks: insect size is limited by the space requirements of their strange respiratory systems, but in the Carboniferous they could breathe in less air to get the same amount of oxygen, and so could reach otherworldly sizes.

When you put a log on the fire, the light and heat you see is, in a literal sense, the decades of sunshine that tree basked in over its lifetime. The solar power is stored in chemical bonds, and the carbon dioxide released from the flames is the same that the tree inhaled to synthesize its sugars and build its wood and leaves. When we retrieve eons-old coal forests and burn them in power plants, we release the millions of years of prehistoric sunlight and carbon dioxide trapped in them. This ancient sunlight warms us in the winter and moves our modern world. But after geological ages of rock-bound slumber, we're now releasing—all at once—the same carbon dioxide responsible for the difference between the tropical greenhouse of the Devonian and the wintry climes of the Late Paleozoic Ice Age that followed. We do so at our peril.

After the death of the last placoderm, it would be 100 million years until the next wholesale slaughter on planet Earth. Making the Ordovician and Devonian catastrophes look like mere dress rehearsals, the devastation of the next mass extinction would bring the planet perhaps as close as it has ever come to losing its pulse altogether.

4

THE END-PERMIAN MASS EXTINCTION

252 Million Years Ago

All earth was but one thought—and that was death.

—Lord Byron, 1816

"So 500 million years is a long time, right?" Stanford paleontologist Jonathan Payne placed a polished slab from the End-Permian* mass extinction on his office table—a block of ancient seafloor from China. The rock accumulated over thousands of years spanning the extinction. The bottom half, from before the extinction, was made of ground-up seashells and plankton—the detritus of a living world. The top half, from after the extinction, was made of microbes and mud. Where these layers abruptly met in the middle was the worst thing that's ever happened in the history of life on earth.

* Also known as the Permian-Triassic or Permo-Triassic mass extinction.

"Five hundred million years is a really, really, *really* long time. And this is the single worst event in the last 500 million years of earth history. So your scenario should not be sort of a bad-day scenario. Whatever happened is presumably about as extreme as Earth's surface conditions have been in the last 500 million years. So this isn't a one-in-a-hundred event, it's not a one-in-a-thousand event, it isn't even a one-in-a-million event. It's closer to a one-in-a-billion event. You want to keep that in mind. Whatever this is, it's the worst that it's ever been."

Before apocalypse struck, it was the Permian period. In the 100 million years since the icy end of the Devonian, the planet had at least the broad sketches of a world we might recognize . . . kind of. At the very least, there were now trees and plants on land, and large beasts that trudged among them. This was a profound break from the world that came before. Though plant and animal life on land might strike us as the default setting for Earth, it was revolutionary for a planet whose continents had been barren for more than 4,000 million years.

The fish that had been timidly crawling onto land in the Devonian had made it by now and had split into two reptile lineages: one that would remain reptiles (and eventually give rise to crocodiles, snakes, turtles, lizards, dinosaurs, and the dinosaurs' popular spin-off, the birds) and another group that would eventually become the mammals.* Surprisingly, it was this latter group that ruled the world of the Permian, while the reptile line mostly waited its turn for world domination. This proto-mammal ruling class was an alternate universe of unfamiliar and rather hideous beasts—a menagerie stocked with lithe, menacing apex predators

* Amphibians never made it fully onto land—even today they still have to return to the water to lay their eggs.

and lumbering rhino-sized plant-eaters that gathered in herds around Pangaean watering holes. The larger members of the reptile line that did thrive were warty, tanklike ogres. It wasn't Earth's most photogenic moment. In the oceans, the reefs that had been destroyed in the Late Devonian were back, but even though there were sharks and fish, this was still very much a primitive biosphere. The reefs had a distinctly Paleozoic flavor, made of whole orders of colonial animals that no longer exist. Trilobites, which had barely limped through the previous mass extinctions, still shuffled along seafloors paved with brachiopods. Even sea scorpions—reduced now to mostly freshwater environments after the near-shore slaughter of the Late Devonian—had endured since their beginnings in the Ordovician.

But by the end of the Permian nearly everything would be dead.

At the end of the Permian, Siberia would turn inside out, burbling lava over millions of square miles and swamping the atmosphere with volcanic gases. One gas in particular stands out as the primary killer in what would become the greatest mass death in earth history. Researchers don't study the worst catastrophe ever purely out of academic, or even morbid, curiosity. The End-Permian mass extinction is the absolute end-member—the worst-case scenario—for what happens when you jam too much carbon dioxide into the atmosphere.

In the middle of the Chihuahuan Desert, 120 miles from El Paso, is a window onto the happier times before the planet was nearly sterilized. It was here, along lonely Route 62, that I pulled over, stood at the bottom of the Permian ocean, and snapped a few photos of a towering white promontory called El Capitan. The cliff marks the highest point in Texas and is the limestone prow of the

Guadalupe Mountains, an ancient barrier reef constructed entirely from sea life. Today it towers above the empty and arid kingdom of West Texas, just as it would have towered above the ocean floor in the Permian, more than a quarter-billion years ago. Behind it is McKittrick Canyon, a surprisingly verdant, maple-lined valley where gigantic hunks of limestone tumbled off the reef face in prehistoric submarine avalanches and settled to the bottom of the shelf slope, where they still reside. I brought along a dog-eared copy of Smithsonian paleontologist Doug Erwin's *Extinction: How Life on Earth Nearly Ended 250 Million Years Ago* as my guide to this vacant corner of Texas.

"At the base of the steep escarpment at McKittrick Creek," Erwin writes, "one is standing on the ancient sea bottom of the Permian Basin looking up toward the reef some 1,200 feet above, just as one could today off the Bahamas or some other modern reef if all the water was removed. Hiking up the Permian Reef Trail in McKittrick Canyon is just like walking (or better, swimming) up the face of the reef as it was millions of years ago."

And so I "swam," step by step, up this reef face in dusty sneakers, imagining myself a squidlike ammonoid* bobbing up the wall in a whorled shell, tentacles extended—unaware of the 97 percent annihilation awaiting my kind at the period's end. The reef was built of vase sponges, horn corals, and colonies of brachiopods and bryozoans all cemented together by encrusting algae. Plated sea lilies—the crinoids—reached out from these walls, straining the seawater, as snails and trilobites wandered coyly in and around this spectacular living rampart looming over the open sea. Today this marine tableau, frozen in limestone, is free to explore for any-

* Named after the Greek and Roman god Ammon (adapted from the Egyptian God Amun), who is depicted with coiling ram horns resembling the ammonoid's shell.

one with a gallon of water, a wide-brimmed hat, and a healthy fear of rattlesnakes.

"Entombed here is the world of the Permian," Erwin writes about Guadalupe, "the very last profusion of life before the extinction."

Gigantic cave systems in the Guadalupe Mountains, like the Carlsbad Caverns in neighboring New Mexico, have been etched out of this ancient barrier reef by groundwater and now provide an inside-out look at the Permian sea world. A few thousand years ago—not long after the first appearance of stone points and the first humans who left them—giant ground sloths that lived in these caves disappeared, along with saber-toothed tigers, rhinoceroses, and mammoths. But this geologically recent, man-made extirpation had nothing on the Paleozoic apocalypse hundreds of millions of years before.

Here in West Texas, a healthy planet was perched above the abyss, about to dive in. By the end of the Permian, virtually everything on the planet would be killed off, and in the wake of the slaughter, life on earth would chart an altogether new course.

Trilobites, the standard-bearer for the Paleozoic era, had managed to barely survive every mass extinction for 300 million years, but finally succumbed to the slaughter at the end of the Permian, ending their spectacular run on the planet. Who knows how rich the inner lives of the trilobites were, but it's an experience of planet Earth that finally ended here in the chaos at the end of the Permian. Crinoids and brachiopods, which make up the fossil tapestry of the Paleozoic, got hit so hard by the End-Permian mass extinction that they never recovered. Blastoids went extinct. The reef-building animals of the Paleozoic, the tabulate and rugose corals, didn't just get hit hard, as in previous reef collapses like the Devonian mass extinction, but went entirely extinct.

In the harrowing aftermath of the Permian, reefs were replaced with piles of microbial slime. These are the stromatolites, those uninspiring mounds of muck from the dreary eons before complex life. They had mostly disappeared since their heyday in the boring billions, but in the wake of the worst mass extinction ever, with the oceans as empty as they had been since the bacterial age, these throwback mounds enjoyed a brief, eerie renaissance smack dab in the middle of the age of animals, hundreds of millions of years out of place. In the literature, this microbial strata is referred to as "anachronistic," and its ubiquitous presence in the fossil record after the die-off is chilling. With animal grazers wiped out and some truly hellish ocean conditions prevailing, the primordial seas of the early earth briefly returned to the deathly quiet world, and these wildly archaic bacterial kingdoms held sway.

Over millions of years, as plankton snowed through the oceans and accumulated on the seafloor—millimeter by millennium—some of it became solid rock, called chert, made of billions of single-celled critters. After the End-Permian mass extinction, there's a "chert gap" in the fossil record as this rock of life all but vanishes. The gap illuminates the truth that life and geology are two descriptions for the same reservoir of raw material. Pull the lever on one and there's a response in the other, and vice versa.

On land, there was the wild world of proto-mammals—creatures that appeared reptilian but also vaguely canine, and others, bovine. Because these beasts might have lacked scales, artists often depict them in skin tones, with mangy, irregular wisps of hair—leaving an unmistakably sickly impression. This twilight zone coterie was roundly destroyed at the end of the Permian, though our ancestor—probably a small weaselly proto-mammal—once again, miraculously, survived somewhere. Insects, normally buffered against major crises, suffered their only die-off ever at the End-Permian mass extinction. The plant world was so obliterated by the catastrophe that rivers, formerly confined in narrow winding channels, ceased meandering and instead began to roll forth in sprawling, braided sandy streams, as they had done during the billions of years before plants were there to anchor their banks. Along with the chert gap in the seas, there's also a "coal gap" on land, as trees disappear from the fossil record for 10 million years after the extinction. The large woody conifers and seed fern trees of the Paleozoic were replaced by pathetic ankle-high weeds—quillworts—that spread out over the smoldering planet.

Unnervingly, at the same time that plants all but disappear, a brief spike in fungus suddenly appears in the rock layers of the mass extinction, possibly from dead things rotting all over the world.

The mass extinction brought about the end of not only the 50-million-year-long Permian period but also the entire Paleozoic era, then in progress since the dawn of animal life. The Paleozoic, characterized by those ancient seas filled with trilobites, brachiopods, and unfamiliar reefs, was as different from the age to come as the age of dinosaurs is from our modern world. Perhaps most disturbing is that, although the Paleozoic era had lasted for hundreds of millions of years—encompassing the Cambrian, Ordovician, Silurian, Devonian, Carboniferous, and Permian periods—it ended (in geological terms) in what was almost a subliminal time frame. Investigating Chinese rocks that record the mass extinction in the Permian ocean, legendary MIT geochronologist Sam Bowring found that the entire nightmare took place over a breathtakingly short period of fewer than 60,000 years. The End-Permian mass extinction marked the end of one venerable planet and, after a harrowing convalescence, the beginning of another.

A few hundred miles north of the Guadalupe Mountains, and a few million years in the future, is the San Rafael Swell of Utah. It's an enchantingly desolate wasteland bisected by Interstate

70, along the longest stretch of the Interstate Highway System without any motorist services. The landscape in this stretch of Utah is forbidding enough to attract NASA researchers looking for insights into Martian landscapes, and its stark sterility seems an apt memorial for the end of the world. Here, millions of years after the largest mass extinction in the history of life, the kaleidoscopic diversity of the former Permian sea world—the same one visible in those Texas reefs—is reduced to a rare shell fragment here and there, if any fossils can be found in these rocks at all. As the Smithsonian's Doug Erwin writes about these barren wastes: "Graduate students often do not want to work on the early Triassic because there are so few species that fieldwork quickly becomes boring."

For most life on earth, a deathly silence passes over the fossil record. But a rare few thrived in the empty landscape of the post-apocalypse: vast shelly pavements made almost exclusively of a single weedy low-oxygen-tolerant clam called *Claraia* comprise post-extinction layers around the planet, from Pakistan to Greenland. By default, the creature acquired a silent world that no one else was left to claim. Dismal and monotonous pavements of the opportunistic mollusk reveal an utterly shattered world—one that would take almost 10 million years to repair itself.

"If there's two events in the history of life, there's the Cambrian Explosion and the End-Permian mass extinction," Stanford's Jonathan Payne told me.

The break between life before and after the End-Permian mass extinction was so blindingly obvious even in 1860 that the natural philosopher John Phillips could explain the radically different world that bloomed from the ashes of the End-Permian only with a second act of divine Creation.

Since the dawn of animal life, the End-Permian mass extinction brought the planet as close to sterilization as it has ever been, dwarfing all other mass extinctions and looming, in the story of life on earth, as the planet's all-is-lost moment.

In 2007, University of Washington paleontologist Peter Ward wrote a book, *Under a Green Sky,* in which he argues that carbon dioxide emissions are not just a regulatory headache for bureaucrats but in fact have also been "the driver of extinction" throughout earth history.

The book, along with recordings of a series of talks I heard Ward give at Princeton in 2006 in which he compared the End-Permian mass extinction to our own modern crises, had a profound effect on me.* Ward's talks—a mix of technical exposition and gallows humor—introduced me to the idea that carbon dioxide–driven global warming is not just being simulated in climate models on government supercomputers, but is an experiment that the earth has already run many times in the deep past. More shocking to me was that global warming might have been implicated in the most extreme die-off ever in the fossil record.

"Is it happening again?" Ward asks in *Under a Green Sky.* "Most of us think so, but there are still so few of us who visit the deep past and compare it to the present and future."

Ward's warnings that we might soon be revisiting the worst chapters in Earth's history were, he said, "powered by rage and sorrow but mostly fear."

After reading his books for years, I finally managed to arrange a lunch with Ward between sessions of the annual meeting of the Geological Society of America. Ward is a surprisingly jolly prophet of doom, with a disarming grin and an unshakable urge to veer off-topic. One quickly finds conversations with him

---

* And probably accounts for the existence of this book you're reading.

jumping from authoritative tangents on the morphology of the nautilus shell to the origin of Chipotle's *E. coli* outbreaks. This sort of manic inquisitiveness has kept him bouncing from continent to continent over the course of a wildly productive career in paleontology—from Antarctica to Palau, from Spain to Haida Gwaii—in pursuit of answers to the biggest questions about the history of life.

Ward's first love wasn't paleontology—it was scuba diving. He was led into the sea by a childhood love of *20,000 Leagues Under the Sea* and, later, by an adolescent envy of Jacques Cousteau and the dashing exploits of the RV *Calypso*.

"I mean, to me those were the heroes," he told me. When he was in college, "the *Calypso* came to Seattle, and I was a dive instructor at the time. I think I was twenty, twenty-one, and we were at this drunken party with these beautiful women. Then these Cousteau guys came in, and in just five, ten minutes, off they went with all the women. When you're, like, twenty-one, what inspires you more than that? I thought, fuck yeah, that's what I want to do with my life."

After spending a lifetime scuba diving off remote atolls strewn across the Pacific and Indian Oceans, Ward has become one of the world's top experts on the nautilus, a gorgeous but bashful animal that bobs along reef walls in a shell that's prized by mathematicians for its geometric elegance. The nautilus is a cephalopod, in the same group as squid, octopuses, and cuttlefish. But unlike those animals, it has a near-useless pair of pinhole cameras for eyes, and chemosensory tentacles that are used less for grasping than for sniffing for food.

"These things are basically just giant noses," Ward said. The nautilus has been around for 200 million years and is the sole survivor of a lineage known as the nautiloids that's even older still. We met the nautiloids in Cincinnati. They made it through all five mass extinctions going back to the Cambrian (including the End-Permian ultra-catastrophe). But today the nautiloids are what's known as a "dead clade walking"—a shadow of their former glory and limping toward extinction. Though they've survived every major mass extinction in the history of animal life, they might have met their match in humanity, which has a tendency to destroy what it loves.

"The bummer is their shells just look too good," Ward said. "They're beautiful."

Some nautilus shells can command up to $200 on eBay, a bounty that proves irresistible to poor Philippine and Indonesian fishermen, and Ward has seen the animals disappear from atoll after atoll over the course of his diving career. "I mean, just anything that's beautiful to humans is out of luck."

It was on one such diving trip in pursuit of these evolutionary relicts in New Caledonia that the course of Ward's life was rerouted by tragedy. When his field assistant passed out at a depth of 200 feet, Ward risked his life by hauling his drowned colleague to the surface without pausing on the way up, as divers must do to ward off potentially fatal episodes of the bends.* But the rescue attempt was in vain, and by the time the two reached the surface, his partner was dead. Today the scars of that diving disaster are both physical—Ward had to replace his hip after it was ravaged by the nitrogen bubbles that came out of solution in his bloodstream during that hurried ascent—and psychological.

"It was such a horrendous death," he said.

Ward wrote about how this personal tragedy affected the course of his career:

"It would turn me away from studying the modern, and away from the sea, toward the landward study of darker things, the

---

* Also known as decompression sickness.

study of mass extinctions themselves, for what better way to understand unexpected, unexplained death than to take its measure in its most sepulchral form?"

Given this morbid fascination, it's no surprise that Ward was inevitably led to the End-Permian. The worst mass extinction ever is also known as The Great Dying.

A decade ago, the End-Permian mass extinction had a familiar culprit. In 2004, a team led by University of California–Santa Barbara geologist Luann Becker claimed to have found a giant crater off the coast of Australia. The find bolstered the case made by her team a few years earlier that, like the calamity that annihilated the dinosaurs at the end of the Cretaceous, this even worse mass extinction at the end of the Permian was also caused by a giant asteroid impact. Still, the case that the Permian killer had a celestial origin was far weaker than that for the death of the dinosaurs. One of the main lines of evidence for the dinosaurs' asteroid was the presence in the extinction layers of iridium, an element that's rare on the surface of Earth but plentiful in space rocks. Many researchers assumed that they would easily find a similar signal at the end of the Permian, but despite an exhaustive worldwide search, no one could find much iridium in the rocks anywhere.

Becker's team claimed, however, to have found a different geochemical signal from beyond. Rather than iridium, Becker found *buckminsterfullerenes,* or "buckyballs," in rock samples from China, Japan, and Hungary that, she claimed, came from outer space. These giant carbon molecules are named for Buckminster Fuller, the eccentric inventor of the geodesic dome, which the lattices of carbon atoms are said to resemble. Becker asserted that helium-3 gas was trapped inside these tiny carbon cages and could only have had an extraterrestrial origin. But when a phalanx of scientists descended on her results, not only could they not reproduce the findings, but the Japanese samples turned out to be from the Triassic period. It was later discovered that buckyballs can't trap helium-3 for more than a million years anyway before it leaks out. As for the supposed crater, impact specialists began to highly doubt that the feature had anything to do with space rocks. Most now suspect it to be an artifact of more mundane earth processes. Before long, both findings—the crater and the buckyballs—fell into disrepute, but not before leaving a persistent residue in the science media.

"While popular science magazines such as *Discover* still promote the press-friendly impact hypothesis for the cause of the Permian extinction, among working scientists this is a rejected hypothesis," Ward writes.

While Ward was on sabbatical in South Africa in 1991, the End-Permian wasn't much on his radar. Instead, he was eager to study a local Cretaceous fossil site rich in ammonites, a prehistoric cousin of the nautilus and one that dominated the oceans for hundreds of millions of years. The ammonites had already brought Ward some measure of professional acclaim. His work documenting the spiral-shelled creatures in the beach cliffs of Zumaia, Spain, showed that the End-Cretaceous mass extinction that wiped out the dinosaurs was disastrously abrupt, with ammonites thriving right up until the (very) bitter end. The discovery helped settle the long and acrimonious debate over whether the Cretaceous world petered out over millions of years or was smote in a geological instant.

When a colleague made it clear to Ward that he wasn't in the mood to share his precious South African ammonite fossil site, Ward turned his attention instead toward a famous group of rocks in the desert that were 200 million years older and rich in the sun-bleached bones of long-forgotten beasts. Ward knew that the rocks hovered somewhere in the fossil record near a colossal extinction, one that dwarfed even the calamity that ended the age of dinosaurs.

"I started asking people, 'So what's the record? How close is it to the extinction boundary? What's the pattern of the extinction?' The usual bullshit. The stuff I'd been doing in Zumaia with ammonites. I was shocked to find out that, in fact, nobody had ever done that! They hadn't really thought about doing that! I was totally shocked by what little interest people had in that mass extinction at the time."

The bones interred in the Karoo Desert of South Africa and elsewhere are from the road not taken by our family tree. This is the forgotten world of the Permian, a land populated by our bizarre and formidable cousins and long overshadowed by the mythic reign of the dinosaurs to follow.

The idea that our relatives ruled the world more than 250 million years ago might come as a surprise to those weaned on the idea that mammals didn't make it big until after the dinosaurs were cleared out by disaster almost 200 million years later. And that's true, as these Permian beasts—called synapsids—were still a long way from being proper mammals. The most famous of them, *Dimetrodon*—a fanged, markedly reptilian-looking beast with a giant sail on its back—is often mistaken by natural history museum visitors as a dinosaur.* But in fact, *Dimetrodon* and its cohort of Permian beasts are our ancient cousins. You are a synapsid, just like *Dimetrodon,* a fact revealed by the peculiar, but analogous, construction of its skull. Other early synapsids include *Cotylorhynchus,* a plant-eater built like a beer barrel that sports a head so comically tiny as to cast the idea of survival of the fittest into disrepute. If these early synapsids—our own extended family members—look

* *Dimetrodon* shares screen time with *Stegosaurus* in Disney's *Fantasia* even though the two were separated by more than 100 million years.

so unfamiliar to us, it is because of the cruel pruning shears of the Permian: a series of extinctions, including the Judgment Day at the period's close, cut this blossoming evolutionary tree down to only a branch or two, including that of our ancestors.

Dimetrodon and its sail-backed friends wouldn't make it to see the Permian apocalypse. They would be wiped out (perhaps mercifully) earlier in the period by a totally mysterious event called Olson's extinction.* But the Permian was largely a synapsid-eat-synapsid world, and these fallen synapsids would be replaced by still more synapsids from our side of the tree. This time it was another hideous group, called the dinocephalians, who took over; these were bulky beasts built like tanks, with some sporting skulls that seemed to explode with strange antlerlike knobs. Eventually, the dinocephalians would be dethroned as well (along with a roster of other graceless synapsids with fittingly weird names like *Moschops*) by yet *another* extinction before the end of the Permian. This extinction event seems to be somewhat less mysterious, coming as it did at around the same time as a mass die-off in the ocean as well as the catastrophic eruption of a huge volcanic province that rent open China—a cataclysm that would have been more than up to the task of trashing the planet. As scientists learn more about this extinction, it's steadily moving up the ranks of worst disasters in earth's history. But even this middle Permian crisis—though severe—was a mere body blow compared to the decapitation awaiting the planet at the period's end.

Despite the calamities that struck within the Permian period, the ecosystem was resilient and recovered quickly. In the moments leading up to the planet's ultimate mass extinction, there

---

* University of Oregon paleontologist Gregory Retallack attributed the extinction to a high-CO_2 greenhouse. Other scientists have disputed whether Olson's extinction even constitutes a real extinction event at all, or is instead an artifact of an incomplete fossil record.

was no indication that the world was about to end. These twilight moments of the Permian belonged to the last great group of Permian mammal forerunners: the therapsids.

The therapsids included dicynodonts, dog- to cow-sized herbivores with giant tusks and beaks that probably trampled the shrubby countryside in herds. In an age before flowers, fruit, or grass, these plant-eaters had to make do in a world that was decidedly short on nutrition. In fact, much of the planet was probably uninhabitable. The ancestor ocean to the Atlantic had been closing since the Ordovician, and by the Permian this marriage of the continents was consummated: the planet's landmasses were reunited, after hundreds of millions of years apart, to form one giant supercontinent stretching from pole to pole. The endless interior of this supercontinent was savagely bleak and arid—a sort of global North Dakota—with unearthly heat and ferocious cold, virtually untouched by rain. This was Pangaea.

So far we've taken for granted the idea that the continents have moved around over geologic time. But this idea—that they drift above an unseen, incandescent conveyor of convecting rock—is one of the most revolutionary ideas in the history of science. Amazingly, it's gained widespread acceptance only about as recently as artificial sweetener. And like most scientific revolutions, it began its life as disreputable, bordering on crazy, speculation.

The theory of continental drift was most famously developed by Alfred Wegener, a German meteorologist whose studies brought him, as most scientific pursuits did at the turn of the twentieth century, to the high Arctic. On expeditions to Greenland, he developed a vision of the continents as similar to the great ice floes that surrounded him: calving apart, drifting and crashing into each other over great expanses of time, and at one point forming a supercontinent in the deep past that he called Pangaea, meaning "all earth." Wegener came to this revelation by making the same observation that most six-year-olds do: that the continents roughly fit together, like puzzle pieces. On top of that, fossils seem to form bands that jump the oceans and connect disparate parts of the world by prehistoric biology. Despite the persuasive case he made, Wegener was roundly dismissed by his contemporaries and didn't live to see his vindication. Like all good Victorian Arctic explorers, he died valiantly on the ice, where he remains today, buried under perhaps 100 feet of snow.

The idea Wegener left behind, continental drift, would eventually upend all of geology. The state of the science before midcentury was not unlike that of astronomy before the conceptual revolutions of Galileo and Copernicus, and explanations of the planet's geological features shared the same tortured logic of Ptolemy's epicycles. But when bathymetric surveys of the seafloor in the late 1950s and early '60s showed gigantic underwater volcanic mountain ranges encircling the world like the seams of a baseball, pushing the continents apart, suddenly everything in geology made sense: volcanoes, earthquakes, island arcs, mountain ranges, deep-sea trenches, the distribution of fossils, and the strange complementary borders of the continents, which were indeed once united in a globe-spanning supercontinent hundreds of millions of years ago—just as Wegener had surmised. This supercontinent, Pangaea, reached its apotheosis in the Permian, when it formed a giant splayed C that stretched from the Arctic to the Antarctic, interrupted in the middle by a titanic east-west mountain range where North America met Africa and South America. The supercontinent was surrounded by a global super-ocean to match, called Panthalassa.

While the rhinolike herbivores munched on their unappealing Pangaean shrubs, the kings and queens of the supercontinent were yet another ancient relative of ours: the menacing gorgonopsids—brawny and vaguely wolflike apex predators with skulls like giant staple removers and teeth longer than those of *T. rex*. These fearsome daggers, which they used to tear the plant-eating dicynodonts limb from limb, included incisors, canines, and post-canines, indicating a lineage inching closer to mammaldom. The gorgonopsids are aptly named for the mythical Greek sisters the Gorgons, who could turn people to stone with their gaze alone.* All these long-lost cousins of ours—dicynodonts and gorgonopsids, herbivore and carnivore alike—ruled the world for the final 10 million years of the Paleozoic, until Armageddon.

It was the dusty bones of these distant relatives that Peter Ward plied to give up their secrets in the End-Permian wastes of South Africa. Grant funding in hand, Ward, along with Roger Smith of the South African Museum, returned to the desert to unpack this worst mass extinction ever. In the Karoo (which was near the South Pole in the Permian), it requires only a short walk to view the startling transition from the epic 100-million-year ice age after the Devonian to the xeric wastes of Permian Pangaea.

"You start out and you see the dropstones, so you've still got ice out there, but then by the end you've gone from—in one period, in one interval of rock—an ice age to the super-hot deserts where things are dying like crazy. It's just a function of a few million years and the whole world reverses."

The first question about the mass extinction at the end of the Permian was the simplest: was it a protracted affair, with the planet wasting away by attrition over millions of years, or was it geologically sudden and catastrophic? It's a question that's surprisingly difficult to answer, and one that required years of collecting skulls

---

* Geologists don't know which kind of stone.

and bones in the Karoo before subjecting the data to the clarifying light of statistics. Ward and Smith found that the mass extinction on land was indeed catastrophic. At what they interpreted to be the boundary between the Permian and the Triassic, the therapsid world all but vanished in what looked to be a timescale of thousands of years, not millions, as had been previously thought. The vicious gorgonopsids were annihilated, disappearing completely, and as for the thirty-five genera of plant-eating dicynodonts known from the late Permian, only two made it through the sieve of mass extinction. In the Karoo, the beginning of the Triassic is announced by the lonesome presence of one of these plucky survivors, *Lystrosaurus,* a deeply unattractive, piglike burrower sporting tusks and a beak to shear the hardy weeds of the wasted world. In artists' representations, *Lystrosaurus* seems to sport the bewildered look of a creature that has inexplicably survived a massacre. In the aftermath of the mass extinction, the unlikely creature inherited the entire earth, dominating the early Triassic fossil record across the globe, from Antarctica to Russia, just like the vast clam monocultures of *Claraia* that pave the seafloors of this post-apocalypse.

Ward, inspired by the Alvarez Asteroid Impact Hypothesis, sought to make a name for himself here in these ominous layers between the reigns of the fallen gorgonopsids and the surviving *Lystrosaurus*. He was after the debris from a catastrophic asteroid collision that could explain the devastation. He hunted for a layer of iridium, bits of fallout ejecta—anything to explain the sudden death of the biosphere. But he couldn't find it.

What Ward and others found instead at the end of the Permian was a wild swing in the carbon cycle.

If the rock hammer is the geologist's best friend in the field, the somewhat bulkier mass spectrometer can be an even more dear collaborator once back in the lab. By vaporizing rock, the machine illuminates the molecular nitty-gritty of any sample. When Ward and his colleague Ken MacLeod subjected chunks of fossil soils, and even *Lystrosaurus* tusks, to this crucible, they found that the amount of isotopically light carbon in their samples skyrocketed at the mass extinction, perhaps reflecting a sudden overabundance of it in the ancient atmosphere. Though the stratigraphy of the Karoo remains the source of ongoing dispute, the results matched findings at End-Permian sites across the planet from the ancient ocean that similarly recorded a carbon cycle jackknifing.

Where did all this extra light carbon in the atmosphere come from? There are a few ways to increase this reservoir. One way is to kill all the plants, plankton, and animals in the world. Plants are picky about their carbon and prefer the isotopically lighter stuff, locking up a vast amount of the world's supply. So too for plankton. And since animals eat those plants, and carnivores eat the animals that eat those plants, the entire living world pulls a huge amount of light carbon out of the system. Therefore, when almost all the plants and animals in the world die, that lighter carbon is no longer locked up in trees and in plankton blooms and animal flesh and there's more of it left over in the atmosphere and oceans. Perhaps, then, this mass death explains the shift in the rocks to lighter carbon isotopes. But the carbon isotope swing at the End-Permian mass extinction is so severe that many other scientists think that the collapse of the biosphere alone isn't enough to explain it.

When the Industrial Revolution began in the eighteenth century and enormous coal measures were ignited in British factories, the world's atmospheric balance of carbon shifted toward isotopically lighter values, reflecting this huge injection of CO_2 from fossil plants. This is another, more straightforward way to get the signal found in the rocks of the End-Permian: simply inject huge amounts of carbon dioxide into the atmosphere.

As Ward said, it doesn't matter whether carbon dioxide comes from "Volvos or volcanoes." At the End-Permian, there were plenty of the latter.

There's simply no modern analog to the eruptions that laid waste to Russia, and the world, 252 million years ago. One tenet of geology, first articulated in the nineteenth century and followed assiduously ever since, is that "the present is the key to the past." This is what's known as "uniformitarianism." It's the idea that we can understand earth history by appealing to geological processes that we see in operation on the planet's surface today. But the doomsday volcanism at the end of the Permian in Siberia refutes this hoary maxim. Like the catastrophic Chinese volcanism earlier in the Permian, the so-called Siberian Traps were an altogether different style of eruption than that with which we're familiar—and occurred on a scale that beggars the imagination. Unlike today's postcard-ready stratovolcanoes in places like Mount Fuji, Vesuvius, or Mount Rainier (or the ones that continually exploded throughout the Ordovician), the Siberian Trap eruptions are what's known as "continental flood basalts." And they are what they sound like: burbling floods of lava that cover whole continents, stacking up miles thick in frighteningly short periods of time (geologically). They are the single most destructive force in the history of animal life. Luckily they don't happen very often.

At the end of the Permian, Siberia briefly turned inside out as the Traps covered Russia in more than 2 million square miles (5 million square kilometers) of lava. Today the Traps comprise soaring plateaus and precipitous river canyons carved out of basalt—landmarks that would be considered wonders of the world were they not concealed in the plain-sight terra incognita of Siberia. Enough lava erupted here at the end of the Permian to cover the contiguous United States in molten rock *a half-mile* deep. In parts of Russia the lava stacks up almost *two and a half miles* deep. The potential detonation of Yellowstone, which would cover some US states in a few inches of ash, is not even worth discussing in the same book as these End-Permian floods of lava.

In 1991, UC Berkeley geochronologist Paul Renne dated the eruption of the Siberian Traps to broadly the same time as the End-Permian extinction, a finding that raised eyebrows in a research community then intoxicated by the idea of asteroid impacts.

These floods of lava gain their deadly bite in a rather unexpected way. It isn't that a font of molten rock simply covers or cremates life on earth. One of the guarantees in biology is the resurrection of life after being smothered by lava. Known as succession, such biological renewal is evident today in the vernal slopes of Mount St. Helens, which in 1980 was reduced to a post-apocalyptic ash heap. And if smothering the continents was sufficient to sterilize them indefinitely, we might not expect today's vast boreal forests of Canada to exist: only a few thousand years ago, the country was suffocated by ice more than a mile deep.

No, the primary kill mechanism of continental flood basalts is the enormous volume of volcanic gases they release, and the most important of these might have been carbon dioxide, which can short-circuit the global climate and wreak havoc with ocean chemistry. And as if the tremendous volumes of carbon dioxide that would have been released from the volcanoes themselves wasn't scary enough, the magma might have erupted through the worst possible place on Earth.

University of Oslo geologist Henrik Svensen has been to the Siberian Traps. It's a voyage that usually requires some combination of planes, cars, helicopters, and, finally, a leisurely float down a river—and off the map. But no matter how much his team tried to get away from it all, there was no telling where hardy Russian bon vivants might show up.

"We were dropped off in the middle of nothing—nowhere, after a two-hour flight in a helicopter," Svensen said about one such trip. "The next day, while we were camping, all of a sudden there was a small, really strange, homemade boat coming down the river, with oil barrels and a wooden plateau on top. It was Russians on vacation! In this area!"

Besides the ancient stacks of lava produced by the Traps, Svensen had also heard about curious pipelike structures in the Siberian subsurface scattered throughout the wilderness. Some pipes were a mile wide, filled with shattered rock; in some places they were capped by enormous craters. These craters, and the pipes beneath them, were afflicted not by impacts from above but by explosive cauldrons brewing far below.

When Svensen went looking for old rock cores that had been drilled by Russians looking for strontium and magnetite ores, he found them languishing in abandoned storage facilities in the forest. Many of these "storage facilities" had since become open-air facilities; missing roofs and walls, they had been burned down or otherwise sacrificed long ago to the Siberian winters.

"We were lucky to find intact cores in these buildings that were completely destroyed," he said. "I'm still working on a lot of interesting material we found in the forest."

The picture Svensen pieced together added new menace to the End-Permian volcanism. When the magma from the Siberian Traps came up through the earth, it intruded into the Tunguska sedimentary basin, a huge swath of Russia that had been accumulating strata for hundreds of millions of years since the Ediacaran period. The basin was filled with carbonates, shales, coals from ancient forests, and enormous layers of salt from bygone dried-up seas. In places these sediments stacked up more than 12 kilometers thick. The Tunguska sedimentary basin is the world's largest coal basin, and it's not a package of rock through which you'd want to send millions of cubic kilometers of lava if you could avoid it. When the magma hit the salt layers, Svensen said, it occasionally got stuck and seeped sideways in giant magmatic sills that ignited the ancient coal, oil, and gas buried under the Permian landscape.

And then—*boom.*

Animals nearby would have witnessed the sudden detonation of the countryside. These were the first salvos of the End-Permian, and they heralded the apocalypse.

The pipes that Svensen investigated were filled with shattered rock as the searing gas rocketed up through the earth and exploded at the surface in cataclysms that left behind half-mile craters.

These spectacular explosions would have supercharged the atmosphere with carbon dioxide and methane, an even more powerful greenhouse gas than carbon dioxide that turns into carbon dioxide when it degrades. It's this fossil fuel combustion, Svensen said, that accounts for the huge crazy swings in carbon isotopes at the extinction—and that even accounts for the extinction itself.

"When you heat the sediments, the carbonates generate CO_2, then the shales generate methane from organic matter, and then the evaporites [salts] in Siberia at that time contained petroleum deposits, like oil and gas, which were all also heated by the intrusive magma."

The cause, then, of the End-Permian mass extinction and our own looming modern catastrophe might have been one and the same. The Siberian Traps intruded through, and cooked, huge stores of coal, oil, and gas that had built up over hundreds of millions of years during the Paleozoic. The magma had no economic motive, but the effect was broadly familiar: it burned through huge reserves of fossil fuel in a few thousand years as surely as fossil fuels ignited in pistons and in power plants.

Svensen's explanation reminded me of a conversation I had had with UC Irvine geoscientist and climate modeler Andy Ridgwell about the modern project of civilization.

"Basically the entire global economy rests on how quickly we can get carbon out of the ground and put it in the atmosphere," Ridgwell told me. "That's basically the global enterprise. And there's a lot of people doing it. Geologically, it's a really impressive effort."

So were the Siberian Traps.

Today humans emit a staggering 40 gigatons of carbon dioxide a year, perhaps the fastest rate of any period in the last 300 million years of earth history—a period that, you'll note, includes the End-Permian mass extinction. Burning every last oily drop and anthracite chunk of fossil fuel on earth would release roughly 5,000 gigatons of carbon to the atmosphere. If we do so, the planet will become unrecognizable, with huge swaths rendered uninhabitably hot for mammals like us (to say nothing of the more than 200 feet of sea level rise that would drown much of civilization).

But as exceptional as humans are, estimates of the carbon released in the End-Permian mass extinction range from an utterly catastrophic *10,000* gigatons of carbon—twice as much as we could ever burn—up to a mind-meltingly unfathomable *48,000* gigatons. As a result, temperature estimates for the End-Permian mass extinction and its aftermath strain belief. In the Karoo Desert, as rivers stopped winding, insects stopped buzzing, and mass death swept over the land, the temperature might have jumped as much as 16 degrees Celsius. On Pangaea, 140-degree-Fahrenheit heat waves wouldn't have been unusual. In the tropics, ocean temperatures skyrocketed from 25 degrees Celsius—similar to today's oceans—to perhaps upwards of 40 degrees Celsius (104 degrees Fahrenheit). This is the temperature of a hot tub, or as End-Permian expert Paul Wignall puts it, that of "very hot soup." Multicellular life simply can't exist in this sort of globe-spanning

Jacuzzi. The complex proteins of life denature—that is, they cook. The language of academic papers is typically measured and sober, but even the peer-reviewed science literature describes the early Triassic period that followed this worst mass extinction ever as a "post-apocalyptic greenhouse."

The devastation loosed by the Siberian Traps wasn't restricted to global warming. When the lava incinerated the mile-thick deposits of salt in the Tunguska basin, this explosive recipe would have yielded a toxic cocktail of horrific chemicals like halogenated butane, methyl bromide, and methyl chloride that, among other things, would have destroyed the ozone layer. Svensen argues that lethal UV-B radiation provides yet another kill mechanism in a world not wanting for executioners.

As further proof, UC Berkeley paleobotanist Cynthia Looy and her colleagues have found strange, malformed spores and pollen grains from plants at the End-Permian, from Italy to Greenland to South Africa, which could be the result of UV-B-induced mutations. I spoke with Looy, who doesn't think that high heat alone would have been sufficient to kill the plant world. "It's really difficult to kill plants," she told me. The abnormal spores and pollen might indicate that radiation levels in an End-Permian world newly stripped of its ozone layer had become intolerable for life on land.

Humanity came surprisingly close to reproducing this doomsday scenario in just the last few decades. The 1989 Montreal Protocol to phase out ozone-destroying chlorofluorocarbons (including End-Permian gases like methyl bromide) is widely acknowledged to be the most successful environmental international agreement ever. But failure was never really an option. NASA simulations of the planet under a business-as-usual emissions scenario for these chemicals showed the ozone layer almost disappearing from the planet entirely by the 2060s, an unimaginable situation that would have doubled UV radiation at the planet's surface and spawned a global wave of lethal mutations and cancers.

International negotiations managed to stave off the prospect of life-threatening radiation by midcentury, but the effort to stanch the hemorrhage of greenhouse gases into the atmosphere has been woefully inadequate, despite similarly alarming computer model simulations of business as usual. This is because the halocarbons covered under the Montreal Protocol (some of the same chemicals that would have boiled out of Russia at the End-Permian) are a rather niche group of industrial chemicals, amenable to global regulation and replaceable by a slew of viable, market-ready alternatives. By contrast, the entire global economy is based on the combustion of fossil fuels, which, unnervingly, might have been the most important component of the End-Permian guillotine. Burning coal, oil, and gas has underwritten the very flourishing of humanity since the Industrial Revolution. As Bill Gates recently told the *Atlantic,* "Our intense energy usage is one and the same as modern civilization."

No one knows where our modern experiment with the planet's geochemistry will lead, but in the End-Permian, massive injections of greenhouse gases into the atmosphere led straight to the cemetery.

That injecting huge amounts of carbon dioxide into the atmosphere rapidly warms the planet is an uncontroversial concept in geosciences and has been fundamental to the field for more than a century. But warming is only one consequence of jacking up CO_2. In addition to heating up the planet, carbon dioxide reacts with seawater to make it more acidic and robs the ocean of carbonate. Since many animals—like corals, plankton, and creatures with shells, like clams and oysters—rely on narrow pH ranges and an abundance of carbonate to build their skeletons, quickly infusing the ocean with a deluge of carbon dioxide can be lethal to them. Today the pH of the modern ocean is falling fast, already by a staggering 30 percent since the start of the Industrial Revolution. Even people unmoved by the galaxy of evidence for global warming have no rebuttal to ocean acidification. It's simple chemistry.

Most frightening for our world, it was ocean acidification that Stanford University paleontologist Jonathan Payne thinks was *the most important* kill mechanism in an End-Permian ocean in which, rounding up, roughly everything died.

Many mass extinction experts are grizzled veterans of the dinosaur extinction wars of the 1980s and '90s, and all the relational and professional fallout of those debates. Payne, though, was still in college by the time the fate of the dinosaurs had been resolved to most people's satisfaction. He's part of a younger wave of paleontologists who increasingly glean the story of life on earth from giant data sets as often as from dusty, far-flung rock exposures. I met with him at his Stanford office, which he occasionally visits between jaunts to China to study the End-Permian mass extinction in the oceans (which was nearly total).

To Payne, the End-Permian hellscape represents the outer bounds of what is possible within our climate and ocean system. It is an absolute worst-case scenario. Nevertheless, it may still prove gloomily relevant for the challenges facing humanity.

But first, it's worth putting the insanity of the End-Permian chaos into context.

Though climate science was long an esoteric field, today a familiarity with the basics should constitute a core part of any responsible civic education for citizens of planet Earth. One number in particular is indispensable to the conversation about the challenges facing humanity in the next few centuries: the amount of carbon dioxide in the atmosphere, as measured in parts per million. For the last few million years, CO_2 levels on the planet have swung between around 200 parts per million during the ice ages and about 280 parts per million during much warmer times. This is where the planet was before the Industrial Revolution and during all of human civilization before it, which took place during a remarkably stable climate window. Environmental activist Bill McKibben started the website 350.org to highlight the fact that beyond 350 parts per million is truly dangerous territory, completely outside of human experience. When the world shockingly hit 400 parts per million in 2013, scientists around the globe reacted with horror.

This global chemistry experiment, if left unchecked, will almost certainly threaten the stability of civilization. The last time carbon dioxide hit 400 parts per million, sea level eventually rose 50 feet higher than today. But 350.org might have needed a few more decimal places at the end of the Permian.

"So taking the modern ocean and adding 40,000 gigatons of carbon—like in the End-Permian—it would take you from, say, 300 ppm to 30,000 ppm CO_2," Payne said.

We both started laughing. This number is incomprehensible. An atmosphere with 30,000 parts per million of CO_2 is no longer planet Earth.

"We don't actually think it went to *30,000* ppm, do we?" I asked.

"We don't really know," Payne said. "I don't know if that's a crazy number or not." He elaborated: "The way I think about it is this: Think about the fact that the End-Cretaceous mass extinction [asteroid] impact produced something like 500,000 times the energy of all the nuclear weapons that have ever been detonated on earth. It created a 200-kilometer crater. The crater would go here, halfway to Los Angeles. It's truly inconceivable. And that didn't affect the biosphere nearly as badly as the End-Permian. So whatever this was, it was very extreme.

"We need something that's going to cause the extinction of 90 percent of the species in the oceans . . . without the help of overfishing," he said, laughing. "So that's important to keep in mind too, right? Most of the extinctions we've observed in the last couple of millennia, most of those are due not to climate change but to direct human interaction—overfishing, overhunting, direct destruction of habitat, not due to climate change or ocean chemistry change. In the End-Permian, you don't have that help. I mean, this has to be *all* climate and ocean chemistry. And so . . . I don't think we have any evidence that can rule out 30,000 ppm CO_2. And if I had to bet you, I would bet that the atmosphere was closer to 30,000 ppm CO_2 than 3,000 ppm."

(For now, typical estimates place End-Permian atmospheric carbon dioxide somewhere around 8,000 parts per million, which is still ludicrously high.)

All that extra carbon dioxide, if injected quickly enough, would not only warm the planet to the sci-fi temperatures discussed earlier but utterly ravage the ocean. The sea would absorb the carbon dioxide and the pH of the ocean would plummet—just as it is starting to do in our own modern ocean.

But timescale is everything. In the long term the oceans can keep up with a huge buildup of carbon dioxide in the atmosphere, as long as it happens slowly enough. The gradual processes of weathering break down rocks on land, washing them into the ocean and, in doing so, buffer the seas against acidification like a Tums to an upset stomach. And the more CO_2 you inject into the atmosphere, the faster the rocks weather away.

"Adding CO_2 increases weathering for two reasons," Payne said. "One is that it makes rainwater more acidic. But the thing that a lot of geochemists think might actually be more important is that it just warms the planet, which creates more evaporation and more runoff, and the more water you pump through the system the more you can drive chemical weathering."

But rock weathering takes time. Lots of time. Like the proverbial bird-that-sharpens-its-beak-on-the-side-of-a-mountain-eventually-chiseling-it-away-to-nothing sort of time. Jack up carbon dioxide in the atmosphere on timescales faster than the rocks can weather away and you have a recipe for ocean acidification.

"So timescale's really important," said Payne. "The response of the ocean to these things depends on timescale. In the long run you add lots of extra carbon into the system, and most carbon comes back out geologically as limestone (calcium carbonate). So in the long run, all this coal we're burning, all this petroleum we're burning, eventually it's going to lead to more limestone in the ocean. The problem is that the timescale for that is 100,000 years. Which doesn't help people. So if you think about the modern ocean, what we're doing is basically just burning CO_2—so we're adding carbon to the ocean—but we're not adding any calcium, right? Nobody's burning calcium and sending it into the atmosphere."

Paleontologists rarely get to see their hypotheses play out in real time, but the modern oceans of the Anthropocene offer something like an unwelcome proof-of-concept for Payne and company. Coral reefs, which today supply one-quarter of the ocean's species, are likely doomed if even modest CO_2 emissions scenarios come to pass, but the bottom of the food chain is already struggling in seas newly awash in anthropogenic carbon dioxide. Today, in the acidifying waters of the Southern Ocean, the shells of small, translucent, fluttering planktonic snails called pteropods, which form part of the base of the Antarctic food chain, have been found pitted with holes. In 2008, NOAA scientist Nina Bednaršek discovered these corroded creatures during a research cruise around Antarctica. By 2050, ocean acidification will render the entire Southern Ocean uninhabitable by pteropods, an ecological catastrophe. Since her original upsetting discovery around Antarctica, Bednaršek has also found disfigured pteropods off Seattle, where they form up to half the diet for juvenile salmon in the Pacific Northwest.

"It's not a question of if pteropods will be dissolving, or if they will be compromised—it is certain they will be," Bednaršek told me.

It gets little attention now, but the prospect of ocean acidification in the next few decades could be truly world-changing.

Though the funhouse numbers of the End-Permian dwarf the total amount of carbon we could ever hope to inject into the system, this doesn't rescue humanity. It's the pace of carbon dioxide emissions, not the absolute volume, it turns out, that's everything. This is the reason why—despite the Hieronymus Bosch–like conditions that prevailed on earth 252 million years ago—Payne and his colleague Matthew Clapham at UC Santa Cruz could publish—with a straight face—a paper with the title "End-Permian Mass Extinction in the Oceans: An Ancient Analog for the Twenty-First Century?"

No one knows what modern coral reefs will look like at the end of the twenty-first century, but if the Great Dying is any guide, it could get grisly.

I asked Payne: where there were once resplendent reefs in the Permian, like the ones that built the Guadalupe Mountains, what would scuba divers see if they were to revisit the planet's reefs at the height of the End-Permian mass extinction?

"You'd probably see a lot of green slime," he said. "There might have been big blooms of jellyfish; it's possible. Hard to know."

I asked him what the worst-case scenario for humanity is.

"I think the worst-case scenario is that we acidify the oceans, we kill all the corals and all the other big animals that live in there, and, yeah, you end up with a slime world."

One of the strangest signatures of the End-Permian mass extinction is the presence of a pigment called isorenieretane in marine sediments all over the world—from Australia to southern China to British Columbia. The pigment is used in photosynthesis by a nasty scum called green sulfur bacteria, which requires a peculiar combination of ocean conditions to thrive: no oxygen, poisonous hydrogen sulfide, and, most important, sunlight. If there's sunlight, that means that these blooms of noxious bacteria were appearing in the shallow seas. But an ocean that's bereft of oxygen all the way up to the top is something of an oceanographic nonstarter. The sea surface constantly mixes with the air, oxygenating the top layer of the ocean through the ceaseless churning of wind and waves.

"Uniformitarianism is completely wrong," said Peter Ward. "It's totally wrong. It misleads us. You can't use the present as the key to the past because there were times in the past that were so radically different we can't even conceptualize them. The fact that you can be in the photic zone (the top sliver of the water column where light is able to penetrate) and, even though you've got atmospheric oxygen, only 2 or 5 or 10 meters down you're in a zero-oxygen ocean? That's so weird. That's radically different.

"The whole question really is," Ward said, "how much is Kump right?"

Lee Kump is the head of geosciences at Penn State. He thinks the planet was not only killed by heatstroke at the end of the Permian but also poison-gassed with hydrogen sulfide. And he also has some odd tips for home decorating.

"You know those lamps you can buy now that are salt blocks for cooking steaks?" he asked me in his office in State College, Pennsylvania. "You should buy some, because they're mostly End-Permian salt deposits."

As the interior of Pangaea was becoming an arid, supercontinental hellhole at the end of the Permian, inland seas around the world were drying up, leaving behind huge (and now economically important) salt deposits. Where I live, in Boston, the wintry roads are de-iced with Permian salt from Ireland.

"I was just out at our grill place, and they're selling these things—Tibetan salt blocks—for cooking on your grill. So I bought one. Plus we have a lamp in our house for a little bit of decoration."

Along with extremely high heat, devastating ocean acidification, and ozone destruction, other proposed Permian killers include: intense forest-killing acid rain from volcanic sulfur dioxide, brief blasts of *cold* from sun-blocking sulfur aerosols, agonizing respiratory death from the noxious slew of gases billowing out of the volcanoes (gases that would not have been unfamiliar on a World War I battlefield), direct carbon dioxide poisoning, and mercury toxicity. With so many potential killers running amok, Doug Erwin has humorously dubbed the glut of suspects at the end of the Permian the "Murder on the Orient Express" theory of mass extinction.

"Only Dante could truly do this world justice," he writes.

"Yeah, so there's certainly no shortage of killers," said Kump.

Add to this whodunit two more suspects—the horrifying specters of ocean anoxia and anoxia's poisonous bedfellow, hydrogen sulfide, which can be produced by bacteria only in the absence of oxygen.

If you've ever smelled rotten eggs, you know what hydrogen sulfide is. At only one part per million, it already starts to suffuse the air with the unmistakable miasma of rank shit. At 700 to 1,000 parts per million, you die instantly. And this happens. Hydrogen sulfide is also known as "manure gas," and when inhaled in high enough concentrations, it's claimed the lives of countless agricultural workers working in manure pits. It's also a hazard around oil and gas wells, like those (poetically) in the Permian Basin in Texas, where drillers have been killed by the gas leaking up from the underlying rocks.

In 2005, Kump proposed that this foul gas might have been responsible for the Great Dying. To get hydrogen sulfide, first you need anoxia—a more than capable killer in its own right. And as in other mass extinctions, the lifeless rocks characteristic of suffocating oceans exist all over the world, from the Salt Range in Pakistan to the Dolomites of northern Italy, to southern China, to the western United States, to Greenland, to the former whaling outpost Spitsbergen in the Arctic Ocean, and beyond. Anoxia in the oceans seems to be a global signal at the end of the Permian. And it didn't fully dissipate for millions of years after the extinction, perhaps explaining the brutally slow recovery.

In an attempt to explain these suffocating seas, scientists originally speculated that heating up the planet—as the huge injection of carbon dioxide from the Siberian Traps must have done—reduced the temperature difference between the poles and the tropics and brought the global ocean circulation to a halt. As I'm writing this, parts of the Arctic just endured a month 16 degrees Celsius warmer than normal, and ocean circulation appears to be slowing down as Greenland quickly melts. If the circulation stopped altogether in the Permian, paleoceanographers speculated, then the deep ocean would lose its oxygen and anaerobic bacteria would take off, suffusing the oceans with hydrogen sulfide.

But subsequent modeling has revealed that it's all but impossible to stop the oceans like this. Undersea volcanoes, regional salinity differences, and the quirks of oceanography eventually kick circulation back into gear, however sluggishly.

"Stagnant oceans can't really exist," said Kump.

This is good news, but how then to explain the signals of global anoxia and hydrogen sulfide?

Kump thinks that the anoxia was driven not by ocean stagnation but by the extreme heat itself. Simple physics dictates that warmer water is able to hold less oxygen. As an unfortunate accident of animal physiology, it's also true that the warmer it gets, the more oxygen animals need to consume, so a lack of oxygen rapidly becomes a problem as the seas start to warm. But another factor driving the anoxia at the end of the Permian might have been, once again, the weathering on land, which—just as in the Devonian crises—was pouring nutrients like phosphorus into the sea like mad, where it fed explosive plankton growth and a sickly suffocation.

"We've simulated environments under greenhouse climates where you have more intense weathering on land and phosphate is delivered to the ocean, which provides nutrients," said Kump. "It's like a polluted pond in a sense.

"But unlike ponds, the ocean also has sulfate, so then you start to generate hydrogen sulfide."

In addition to the ominous presence of green sulfur bacteria in the ocean, microscopic beads of fool's gold (pyrite) are found at End-Permian marine outcrops around the world, a telltale sign of a water body suffused with poisonous hydrogen sulfide.

But Kump's conceptual leap is that, not only would the hydrogen sulfide have killed any animal in the ocean it came in contact with, but it might have also been responsible for the mass death on land as well. In 2005 he wrote a paper arguing that huge noxious bubbles of hydrogen sulfide might have come to the surface and wafted out of the sea, spreading out over the land, covering the earth in a toxic, putrid haze, and killing almost everything.

When subsequent computer modeling showed that such a catastrophic gas release from the ocean was unlikely, Kump had to shelve the idea, and other kill mechanisms stepped to the fore. But he hasn't abandoned the nightmare scenario altogether.

"So let me tell you about my latest horror movie concept," he said. "The people out at the National Center for Atmospheric Research [NCAR] have these fancy models that can run through the daily cycle and through the annual cycle, and with Permian conditions they've been generating hypercanes."

Uh-oh.

Hypercanes are continent-sized hurricanes-from-hell, with 500-mile-per-hour winds, that surprisingly pop up in atmospheric models whenever ocean temperatures are ramped up into uncharted territory. Like seawater hotter than 100 degrees Fahrenheit, 500-mile-per-hour winds are almost inconceivable. They're 200 miles per hour faster than the fastest winds inside of the strongest tornadoes. They're the sort of wind speeds that are only briefly achieved directly under nuclear blasts.

"These are mega-hurricanes, which can penetrate clear up to the Arctic Circle—hugely powerful. They cover, you know, the entire continent. So they're so huge—just incredible expanses—and they have great power to penetrate onto land. And so, one of the things I've been trying to get the people at the NCAR to do is to have one of those cross an ocean in which there's hydrogen sulfide, because the hypercanes are going to suck it up."

The horror movie was coming into focus.

"So you'd have these hurricanes, that not only have these 500-mile-per-hour winds, but that are loaded with hydrogen sulfide"—he started laughing—"*and* carbon dioxide. So you'd have these poisonous hyper-hurricanes coming across the land."

Kump kept laughing. It was the nervous laugh of someone who

knew he was saying something insane and terrifying . . . and possibly true.

To summarize: There was an ocean that was rapidly acidifying—one that, over huge swaths of the planet, was as hot as a Jacuzzi and completely bereft of oxygen. There were sickly tides suffused with so much carbon dioxide and hydrogen sulfide that either poison would have sufficed as a killer in its own right. There was a Russian landscape detonating and being smothered in lava several miles deep. There was a fog of neurotoxins and lethal smog streaming from these volcanoes and, high above, an ozone layer blasted apart by halocarbons, inviting a bath of lethal radiation at the planet's surface. There was forest-destroying acid rain and a landscape so barren that rivers had stopped winding. There were carbon dioxide levels so high, and global warming so intense, that much of the earth had become too hot even for insects. And now there were Kump's unearthly mega-hurricanes, made of poison swamp gas, that would have towered into the heavens and obliterated whole continents.

Given how outrageous some of these End-Permian scenarios were, I asked Kump whether comparisons to the modern day are really appropriate.

"Well, the rate at which we're injecting CO_2 into the atmosphere today, according to our best estimates, is ten times faster than it was during the End-Permian. And rates matter. So today we're creating a very difficult environment for life to adapt, and we're imposing that change maybe ten times faster than the worst events in Earth's history."

"That's the take-home message."

He chuckled again.

"Not to be a gloom-and-doom guy."

5

THE END-TRIASSIC MASS EXTINCTION

201 Million Years Ago

It's almost cheerful news that there would be still more mass extinctions in Earth's future. For anything unlucky enough to witness the unimaginable crescendo of the End-Permian, it must have seemed certain that it would be the planet's last.

But failing the complete sterilization of every last lagoon, cave, secluded pond, and deep-sea canyon on earth of even their most weedy and uninspiring inhabitants, the planet can survive. Indeed, in the wake of the major mass extinctions it does more than survive: it blooms anew, which is what it eventually did (quite literally) in the Triassic. Tens of millions of years after the nadir of earth history, the battle-weary supercontinent blossomed again and now played host to the mythic age of reptiles. But the good times didn't last long. As happened at the end of the Permian, the earth would open up once more at the end of the Triassic and swallow the biosphere.

Time is exceptionally cruel to preservation, making the mere existence of the fossil record something of a miracle. Much of earth history has been erased, churned up, and obliterated by the ages. But this isn't the case for the 200-million-year-old planet killer of the Triassic, which suffers no such obscurity. The culprit for the End-Triassic mass extinction, which wiped out three-quarters of life on earth, is still strikingly visible from almost any building on the west side of Manhattan.

But to get to a mass extinction, first you need things to kill, and before the world could be destroyed again, it had to recover from the worst thing that ever happened. This wasn't easy. Though a new, confident world would be established by the end of the Triassic period, at the beginning of the period the planet was still ruined beyond all recognition and scarcely inhabitable. It must have seemed, even after the peak of the End-Permian catastrophe, that these were the last miserable days of planet Earth. Where life existed, it was dominated by invasive disaster opportunists like the ubiquitous clam *Claraia,* while things like trees remained curiously absent for 10 *million* years. It was once thought that this extremely delayed recovery was a product of the unprecedented magnitude of the End-Permian apocalypse. If you punch someone in the face, it might take some time for that person to hobble back up. But if you hit someone with your car at 100 miles an hour, it will be even more difficult to coax them back to their feet.

More recent work has shown, however, that it might not have just been the intensity of the End-Permian extinction that kept the earth foundering in the aftermath of the Great Dying, but the relentlessly bleak and otherworldly conditions that persisted well into the Triassic. Recent science papers don't mince words about this hellish planet: "Lethally Hot Temperatures During the Early

Triassic Greenhouse" announced one paper in the journal *Science* in 2012. The study, by China University of Geosciences geologist Yadong Sun and her colleagues, analyzed oxygen isotopes in the fossil teeth of tiny eel-like creatures to show that sea surface temperatures approaching 40 degrees Celsius (104 degrees Fahrenheit) persisted in the tropics as much of the ocean remained inimical to life for millions of years. On land the planet's entire lifeless midsection endured otherworldly temperatures upwards of 60 degrees Celsius (140 degrees Fahrenheit). This extreme heat corresponds to a lack of large fish fossils over the entire middle of the planet in the early Triassic and a similar lack of animals anywhere near the tropics on land. What recovery did happen, like the surprising evolution of dolphinlike reptiles called ichthyosaurs, was mostly relegated to the poles. Analyzing uranium isotopes in the rocks, Stanford's Jonathan Payne showed that anoxia in the oceans also remained a chronic stressor for 5 *million* years after the extinction. Cruelly, there was even another major pulse of extinction among the few survivors of the End-Permian only 2 million years after the dust settled on the Great Dying.

Perhaps it's less than a coincidence that the most persistently awful time in the history of complex life happened to be the only time the landmasses were united as a supercontinent. The unusual configuration of Pangaea might have broken the planet's thermostat by robbing it of its ability to regulate its atmospheric carbon dioxide. Though the edges of the supercontinent weathered, drawing down CO_2, the vast dry interiors saw virtually no water at all. No water meant no weathering, and no weathering meant that the earth's most reliable mechanism for drawing down CO_2 was broken.

"So when you form a supercontinent in our climate models, you end up with dry interiors," Lee Kump told me. "And so they're kind of a noncontributor to the global carbon cycle at that point because there's no water to weather the rocks, and so, yeah, you could imagine that volcanic eruptions at a time of high continentality like Pangaea would break the regulator for CO_2. Suddenly you have an unabated increase in carbon dioxide."

As a result, the early Triassic was excruciatingly hot.

Some of the other main dumps for CO_2 are in reefs and shallow ocean shelves, where corals (or, in the wake of the extinction, microbes) lock up carbon as limestone, while carbon-rich plankton sinks to the bottom of the sea and eventually becomes rock. It's a simple fact of geometry that having a bunch of little continents gives you more coastline than having one big supercontinent. And more coastline gives you more shelf space to bury carbon in shallow sea life. But in the Permian and Triassic, this space was in short supply around the bloated supercontinent, and simple geometry jammed the biological carbon pump. As a result, more CO_2 gathered in the atmosphere and the planet couldn't manage to cool off. Add to this the enormous CO_2 sinks of trees and forests that nearly disappeared for 10 million years after the Permian and there was nowhere to shuttle all that extra carbon dioxide.

Eventually, though, the planet would cool off, however slowly, and life would fitfully recover. But the early Triassic earth remained a largely broken world, and the wastelands of tropical Pangaea barren and lifeless.

And then, 20 million years after the Great Dying, something beautiful happened. It started raining.

And it rained, and it rained. And it kept raining.

Dinosaurs appeared. Not long after, the first flower blossomed.* Crocodile ancestors followed, along with the first true mammals. The planetary deluge was what's known as the Carnian Pluvial

* *Sanmiguelia,* though its status as the first flowering plant is disputed. Flowering plants wouldn't truly flourish for more than 100 million years.

Event, a little studied but extraordinary event in Earth's history when the floodgates opened and the arid world got a sorely needed soak. It's been called "the greening of Triassic Earth."

But this "greening" might not have been all that benign. In fact, it might have even been accompanied by another minor mass extinction, as many lumbering reptiles and End-Permian stragglers on land seem to have disappeared, making way for the new world. In the seas, slender marine reptiles called thalattosaurs disappeared at the event, while ammonoids once again were hit hard. (Although this isn't a huge surprise: "You look at 'em wrong and they go extinct," University of Chicago paleontologist David Jablonski told me about the volatile boom-and-bust cephalopod group.) The dramatic change in the climate seems to have been kicked off somehow by yet another, smaller flood basalt, which erupted under the Triassic ocean and can be found today in the coastal mountain ranges of British Columbia. A subtle northward shift by Pangaea might have also primed the planet for mega-monsoons.

By the late Triassic, a new order had been established. There were gigantic shovel-headed amphibians basking on the banks of swampy floodplains and hauling out on fallen waterlogged cycads. By now, turtles had arrived on the scene, as had some small flying pterosaurs. And of course, there were new characters dashing through the forests on two legs: the dinosaurs, though they were mostly small and rare. It wasn't their turn yet. Neither was it time for the fallen synapsids and their new family members, the mammals, which would have to wait more than 100 million years for another shot at the top.

The Triassic world, instead, was dominated by a bloodline that survives to this day. This deposed royal family still haunts swamp margins and irritably moseys across golf courses, but in the Triassic world, crocodile kin ruled the earth.

I bounced along a dirt road on the Virginia–North Carolina border, past a sign warning DANGER ACTIVE MINE KEEP OUT, in search of this new world created by the Triassic. I followed a pickup truck, badly rusting and dented from a lifetime of abuse by the Virginia Museum of Natural History. We briefly swerved off the road to avoid a 3-foot-tall theropod dinosaur (aka wild turkey)—which frantically skipped in front of the truck in a flurry of feathers—before pulling into the scale house of the Solite Quarry. Outside the building was a chunk of rock excavated from the site that bore the mark of unassuming three-toed dinosaur footprints (not unlike the turkey's), announcing the arrival of the superstars of the fossil record and their humble beginnings here in the Triassic.

The site had recently switched ownership to creationists, and the Virginia Museum of Natural History was scrambling to recover what was left of this world-famous, 225-million-year-old *Lagerstätte* before its new landlords—unmoved by the old-earth implications of the museum's work—blew up the rocks and turned them into road material. A few weeks earlier, a dig volunteer had driven a car to the site with a COEXIST bumper sticker featuring symbols from all the world's major religions. It infuriated the mine's evangelical owners and nearly cost the museum its access to the site.

"We feel extremely fortunate that they're even working with us, but there's just been a lot of stress associated with, you know, when are they going to blast? When are they going to blast?" said Joe Keiper, the executive director of the museum, as he surveyed the industrial landscape in a hard hat.

"I'm a little nervous, because in this area behind us they've been working on clearing out all the debris in the last few weeks," he said as hydraulic monsters menacingly rumbled by the dig site.

"It suggests that they're prepping this site. We feel fortunate just having one more day, one more day, one more day out here."

Keiper spent the morning excavating a coelacanth from the rocks while I and two other paleontologists peeled back thousands of years of ancient lake bottom layers with rock hammers and chisels, revealing plants, tiny freshwater shrimp by the thousands, and the occasional foot-long swimming reptile. A few years earlier, the museum had discovered *Mecistotrachelos* in the quarry: this truly bizarre little reptile had strange leathery wings under its arms that were built from its ribs and were splayed out and connected by webbing.* It probably glided through the air hunting for bugs on what had once been a Triassic lakeside retreat. And there was plenty for it to eat. Under 2 feet of this slate was the so-called insect layer, one of the most exquisitely preserved in the world. It offered a glimpse into the recovery of the bug world millions of years after its devastation at the end of the Permian.

"They've been underground for 225 million years, but when you get these bugs back to the lab, you can count the antenna segments, and you can count the hairs on the antenna segments—the preservation is miraculous," Keiper said.

In the late Triassic, placid rift valley lakes like these stretched from North Carolina and Virginia up through New Jersey and New York City, all the way to Connecticut and even Nova Scotia. The region resembled today's narrow East African rift valley, where Lake Tanganyika and Lake Malawi have settled in the seams where East Africa is pulling away from the rest of the continent. This is where the "rift" comes from in "rift valley." In the Triassic, the lakes that lined the eastern seaboard and the west coast of Africa were like a perforation running down the center of Pangaea, and it was along this line that the supercontinent would eventually

---

* Like the modern "flying dragon" lizards of Southeast Asia.

tear apart. That's why the eastern seashore of the United States and the west coast of Africa, which share these lake fossils, are where they are. As the supercontinent began to tear apart, water flowed into these great rift valleys, creating the lakes and inviting a strange new world of unfamiliar crocodile cousins to the shores of this tropical Triassic refuge.

Most of these animals would not be recognizable to modern eyes as crocodiles. That's because they're not crocodiles. It would be like calling dinosaurs "birds"—it gets the relationship backwards. Yes, some lumbered on all fours with tapering teeth-filled snouts like modern crocs, but others, like New Mexico's *Effigia,* were swift, lithe, and toothless (!), galloped on two legs, and sported a pair of near-useless stubby forearms. Others, like *Postosuchus,* could be matched up against the over-large velociraptors of *Jurassic Park* and be expected to hold their own. Still others were decked out in fanciful armor, like *Desmatosuchus,* a pig-nosed, almost armadillo-like plant-eater encased in spiky armor with arresting hornlike appendages projecting out of its shoulders; this animal patrolled the floodplains and rivers of the Texas Panhandle. It's some wonder that these animals haven't invaded the daydreams of six-year-olds with the same insistence as *Stegosaurus.*

Although the Triassic croc cousins that ruled the world have long been overshadowed—like all else in the fossil record—by the dinosaurs they dominated, they're beginning to get their due. In the week before our dig at the quarry, another group of paleontologists working nearby, a team from North Carolina State University, had announced the discovery of the so-called Carolina Butcher, a 9-foot croc relative that walked on its hind legs and terrorized these tropical lake margins as the top carnivore of the Carolinas. Artists' depictions of the animal are terrifying: a distinctly crocodilian-looking beast lurches forward, its maw agape in a murderous howl. Another croc relative new to science—this one covered in armor and sporting an intimidating spike-studded collar—had been recently dug out near Raleigh. All these creatures gathered around this vast ancient lake system, which was then marooned about as far from the ocean as you could get in what had been—until the world started pulling apart—the abysmal continental interior of Pangaea. This might have been Pangaean flyover country, but a new feature of the planet was opening up underneath them: the Atlantic Ocean.

As the kingdom of Pangaea finally started to break apart at the end of the Triassic, the world nearly ended once more.

As a teenager, Manhattan-born and Jersey-raised Columbia University paleontologist Paul Olsen explored the banks of the Hudson River beneath the looming cliffs of the Palisades. The towering ramparts of basalt offered a majestic reply in stone to the skyscrapers across the river. Today they're part of a network of monumental seams of magma that suddenly appear across the Atlantic, and on both sides of the equator, at the end of the Triassic. Below the Palisades are still more of the peaceful rift valley lake deposits (just like those in North Carolina) from earlier in the period. Here, with the steady drone of traffic humming from the steel girders of the George Washington Bridge, Olsen would hunt for the relics of a lost world. In the shadow of these cliffs, the self-taught rockhound pried out the remains of ancient reptiles and fish from the metropolitan banks.

In 1970, Olsen's precocious paleontology was beginning to get the high schooler noticed. A successful letter-writing campaign to President Richard Nixon to protect a fossil-rich, abandoned quarry near his house in New Jersey from development earned him a spread in the pages of *Life* magazine at the age of seventeen. The designation of the quarry as a landmark—now known as Walter Kidde Dinosaur Park—came over the objection of some of Nixon's advisers, who urged the president not to engage with the teenager. "Too obvious a joke about 'the Neanderthal Wing' etc.," William Safire and Pat Buchanan scrawled in a memo to the president.

Four decades later, Olsen sports a shock of white hair and a mustache to match, but still retains the restlessness of his adolescence. And his gift for finding unlikely fossil sites never waned. (The fossil riches of the Solite Quarry that I visited on the Virginia–North Carolina border was his discovery as well.) In dealing with dizzying timescales and mass extinctions, Olsen approaches his work with a lighthearted touch. He happily demonstrates the timescale of Earth's history with a pint of beer—with animal life represented only by the foamy head.

Earlier in his career, Olsen, like other paleontologists, wasn't even sure there had been an extinction at the end of the Triassic, so blurry was the fossil record. If the Triassic croc world did lose out, many paleontologists thought, it had been a fair game and the dinosaurs—who would come to dominate in the ensuing ages—just played it better. But Olsen was finally persuaded of the reality of the extinction after a growing body of literature had testified, not only to a massive loss of species on land and at sea, but to a devastatingly abrupt loss at that. Like the mammals that would inherit the earth 135 million years later, the dinosaurs first needed the incumbents—in this case the croc clan—to be violently overthrown in the End-Triassic mayhem before they could take over the world.

Given the apparent rapidity of the mass extinction, Olsen published a number of papers in the 1990s and early 2000s proposing a fashionable culprit: death from above. And there was a main suspect: a 62-mile-wide, nearly perfect circular system of lakes in Manicouagan, Quebec, visible from the International Space Station. The crater was indeed caused by a cataclysmic asteroid collision, but it would later be revealed to have struck 14 million years before the Triassic mass extinction, during a period of relative peace. The discovery that an asteroid not much smaller than the one that wiped out the dinosaurs could have had virtually no impact on life on earth was a shock to a generation of paleontologists raised in the long shadow of the Alvarez Asteroid Impact Hypothesis of mass extinction—the once-scandalous suggestion that life on earth could be wiped out, not gradually over geological timescales, but within minutes from above.

"Here was a large crater, which . . . had earlier been estimated should have been caused by an asteroid large enough to kill off between a quarter and a third of all species on earth, and we found nothing!" writes Peter Ward. "Nothing happened! The lethality of asteroid impacts might have been overestimated."

Olsen began to look elsewhere for the Triassic grim reaper. Meanwhile, at his office at the Lamont-Doherty Earth Observatory in Palisades, New York—above the very cliffs he had explored when he was young—he had literally been sitting on top of his elusive planet-killer the whole time.

For millions of years in Triassic New York City, life followed the unhurried rhythm of a planet in its dreamy, reptilian adolescence—with no hint of the troubles ahead. Twenty-foot, croclike *rutiodons* slid into the water from Newark armed with long, nimble snouts that they used like chopsticks to pick off tropical lake fish and freshwater sharks. They would surge onto the muddy banks of Morocco for an afternoon respite, sending a flock of skittish dinosaurs scurrying through the lakeside horsetails on two legs. Not all the animals would defer as easily. Oversized, ill-tempered amphibians with broad, flat skillets for heads stood their ground, reluctantly consenting to share the shoreline after noting their irritation with a few guttural moans. As twilight fell, tiny reptiles with wings sprouting from beneath their arms would leap and glide from lakeside cycads into billowing swarms of insects that rose from the swampy margins. As the sun set over the rugged peaks of the New Jersey highlands, the deafening stridulations of cricketlike bugs the size of bread loaves would thrum in the conifer cathedrals and echo over the water.

The crust beneath this tableau was thinning like pulled taffy as Pangaea cleaved apart. A gigantic plastic blob of Earth's mantle was surging to the surface on an inevitable trajectory to kill most animal life on the planet. After more than 30 million years of Pangaea pulling apart without incident, something was about to go dreadfully wrong.

When I met Olsen at his office, he shuffled me into his rusting Toyota pickup and hit the gas. While he works in a field that can render a million years utterly inconsequential, he drives as if time itself were running out.

"This car owes me nothing," he said as we aggressively overtook cars on the Palisades Parkway. We parked at the base of the cliffs. In the distance, cranes busily plied the upper reaches of One World Trade Center while a muffled hum wafted over the river. Here, though, separated from the skyline by the Hudson, it was peaceful. Before us towered a neglected edifice of basalt, obscured by invasive ailanthus branches and sumac as well as fading spray-paint professions of love to Jessica and of allegiance to adolescent gangs. I had seen this towering wall of volcanic rock from the other side of the river a hundred times before while visiting New York City. It had always impressed me from afar, with the same dull, superficial awe anyone feels in the presence of big landscapes. But the ghost stories of geology, rather than explaining away the scenery's grandeur, add a vertiginous beauty that multiplies its power and colors the cliffs with an almost menacing indifference to the hives of humanity that buzz all around.

"People are always amazed to find out there's something of this global import right next to the city," Olsen said of the gigantic cliffs that line the Hudson like the rim of a bathtub. Over the decades, the Palisades have attracted the likes of Hudson River School painters and salivating developers, but now are hallowed ground in the study of mass extinctions.

Few people without National Science Foundation grants (or rickety homemade Russian houseboats) will ever see the most dramatic exposures of the Siberian Traps that destroyed the Permian world. Even fewer will see the crater that wiped out the dinosaurs, hidden as it is under millions of years of marine limestone in Mexico. But the continental flood basalt that wiped out the Triassic world isn't remote or obscured; it's a hot spot for real estate developers. So much so that four governors of New Jersey recently wrote a *New York Times* op-ed, "The Threat to the Palisades," about the rampant suburban sprawl taking over these volcanic cliffs. (If they had been alive 201 million years earlier, the op-ed would have undoubtedly read "The Threat *of* the Palisades.")

The cliffs were once gigantic underground channels of magma that spewed their incandescent fountain a little farther west and piled up to create today's Watchung Mountains in northern New Jersey, which sweep out over the state in thick concentric waves of basalt. Turn on the topography on Google Maps and have a look. It's almost exactly as you would imagine a burbling font of lava oozing over the land. Today these piles of ancient lava are green and mottled with suburban subdivisions, evident only to the I-80-bound driver as a sloping terrain that, near Paterson, New Jersey, casts shadows over the big-box stores, office parks, and parking lots below. When the eruptions ended, the underground volcanic plumbing that fed these massive eruptions froze, and in the Palisades they eventually tilted up and eroded, revealing the titanic scale of the End-Triassic volcanism to anyone who knows what they're looking at. These and other eruptions once covered the rending supercontinent with lava over an area equivalent to one-third of the surface of the moon. Known as the Central Atlantic Magmatic Province, or CAMP, it's the Triassic's answer to the Siberian Traps. Landscapes similar to the Palisades from this volcanism exist as far afield as France, Brazil, and Morocco, which was once contiguous with New Jersey and today features soaring sections of the same basalt stacked up in the North African Atlas Mountains.

In 2013, Olsen and a team led by Terrence Blackburn, then an MIT PhD candidate, definitively dated the creation of these picturesque cliffs of Olsen's fossil-hunting youth to the time of the End-Triassic mass extinction. Analyzing rock cores from Morocco and from the Bay of Fundy, along with one core taken from underneath the congested snarl of the highways branching off the George Washington Bridge across from New York City, Olsen determined that, not only was the extinction contemporaneous with the CAMP eruptions, but it was, in geological terms, almost instantaneous. Using unprecedentedly precise radiometric dating, his team determined that the earth first opened up 201.56 million years ago—exactly the time of the global extinctions. The continental flood basalt then erupted in four brief pulses over 600,000 years.

Olsen creatively applied his knowledge of astrophysics to the fossil record to resolve this catastrophe even further. Over thousands of years, the North Star, that unchanging fixture of the sky, does in fact give way to new North Stars as the planet wobbles imperceptibly on its axis. As the planet slowly teeters, the amount of sunlight reaching different parts of the planet changes. For locales near the tropics, the effect can be switching from a monsoon climate to a drier one. As a result, lakes get deeper and shallower in roughly 20,000-year intervals, over and over and over again.

The rocks when the lakes are shallow—red mudstone, with animal footprints and tree roots—are very different from when the lakes are deep—black, thinly laminated, with exquisitely preserved fish fossils.

"The lake sediments are like a rain gauge that's color-coded," Olsen said.

Sedimentary rocks laid down in these rift valley lakes of the End-Triassic are a veritable seersucker of red and black, testifying to the planet's regular wobble.

Olsen determined that the first, most devastating wave of extinction happened within just one of these layers, perhaps in fewer than 20,000 years—a geological instant. Barring the invention of time travel, this is just about as high-resolution as windows into deep time get for geologists. It was an event that, in a staggeringly short time, wiped out three-quarters of animal life on earth, ended the Triassic, and swiftly deposed the ancient crocodile line, cutting short their brief reign.

The unimaginable scale of the volcanism at the end of the Triassic started to sink in after my field trip to the Palisades with Olsen. I began to see this ever-present basalt everywhere I looked. Driving through New Haven, Connecticut, I noticed that the steep treeless face of East Rock, which looms over the city, looks an awful lot like basalt. Indeed it is, and unsurprisingly, it's from around the Triassic-Jurassic boundary. On a North Atlantic right whale survey in the Bay of Fundy in Canada, I was ostensibly there to report on the goings-on in the sea, but I couldn't help marveling as we puttered past the towering cliffs of Grand Manan Island, whose colossal and precipitous profile struck me as almost identical to the Palisades. Sure enough, when I got home, a quick Google search revealed that these gigantic cliffs were created by magma 200 million years ago. In Gettysburg, Pennsylvania, the major features of that historic battlefield—and crucially, the course of the decisive battle itself—were shaped by the apocalyptic geology of the mass extinction. The gradual slope up Cemetery Ridge where Pickett's charge met its grisly fate is shaped by the underlying magmatic plumbing of End-Triassic volcanism: it is giant sills of basalt that give the battlefield its contours.

Then there's the pile of ancient magma that is Little Round Top, where Union colonel Joshua Chamberlain held off a Confederate assault as snipers hunkered down in the End-Triassic basalt playground of Devil's Den 500 yards away. Crisscrossing the battlefield is a network of stone walls that were cobbled together from the boulders of magma and, on July 3, 1863, were draped with bullet-riddled soldiers. Walking along the railroad tracks that cut through McPherson Ridge on the northern side of the battleground, one can view the peaceful world before it was upended by End-Triassic volcanism in the tranquil lake sediments of the sort that are scattered up and down the eastern seaboard. And in the bridge over Plum Run Creek—nicknamed "Bloody Run" after it ran red during the second day of fighting—are the humble footprints, no larger than your hand, of Triassic dinosaurs, pressed into sandstone blocks quarried from the area. The volcanic rock's pervasiveness is a feature not only of the northeastern United States but of North Africa, Europe, and the Amazon as well. In total, the continental flood basalt at the end of the Triassic today covers an area of more than 4 million square miles.

"We're talking about planetary-scale volcanism," Olsen said.

The placid lakeside tableau in Newark would have been transformed into a lake of fire at the end of the Triassic as the earth tore open and filled the valleys with lava. Geysers of liquid rock spurted as high as a mile into the air along fissures in the earth that stretched for hundreds of miles—from Long Island Sound to Quebec, Mauritania to Morocco, and running almost 200 miles underneath the Amazon—leaving behind a smoldering wasteland of black rock.

But as with the End-Permian mass extinction, it wasn't this regional chaos—however extreme—that laid waste to the planet, but the volcanic gases released during the tectonic mayhem.

"One of things we see associated with the extinction is a very dramatic increase in carbon dioxide," said Olsen. Here we go again.

Fossil plant life attests to the skyrocketing CO_2. Plants breathe in carbon dioxide through tiny pores on their leaf surfaces. But there's a trade-off for having too many pores and breathing easier: it's also easier to dry out and die. This is why plants keep pores to a minimum: just enough to breathe, but no more than necessary. In times of high carbon dioxide, they're able to get by with fewer pores as they sip from the CO_2-rich air. In 200-million-year-old fossil plants, University College Dublin paleobotanist Jennifer McElwain found that the number of pores on the ancient leaves plummeted over the End-Triassic mass extinction to accommodate what must have been a deluge of volcanic carbon dioxide. As with the End-Permian, there's also a huge shift in the carbon isotope record over the mass extinction, similarly pointing to a massive influx of carbon to the atmosphere.

"The timescale of the CO_2 is just horrible, absolutely horrible," said Olsen. "We know that it doubled, maybe tripled. We think that each doubling of carbon dioxide is about 3 degrees [Celsius] of temperature change on average—which doesn't sound like much, but that's the difference between the glacial ages and today. It's a significant amount, and it changes the extremes substantially. It probably is not a coincidence that in Death Valley yesterday we had the hottest temperature ever recorded in June in North America."

In other words, for the creatures of the Triassic, 3 degrees might have been the difference between life and death on what was already a warm planet. For context, the International Panel on Climate Change's business-as-usual scenario for CO_2 emissions predicts upwards of 5 degrees of warming by the end of this century.

In the mass extinction layers—like those found in rocks behind a retirement home in Clifton, New Jersey—Olsen and other paleontologists found ancient plant remains and even pollen grains that revealed a plant world rocked by these climatic shocks. It might come as a surprise that something as ephemeral as pollen can survive for hundreds of millions of years, but pollen is in fact one of the most durable biological structures on earth. As paleobotanist Alan Graham writes, "If a hammer, bicycle chain, pair of pliers, and pollen were placed in a platinum crucible and warmed with hydrofluoric acid for a week, the metal objects would be digested or highly corroded, while the pollen walls would remain virtually unaltered."

"One of the things that happens at the time of the mass extinction is that for the plant assemblages in the tropics, the diversity is absolutely decimated," Olsen said.

When the diverse tropical plant world was blown apart in Triassic New Jersey, it was suddenly replaced and dominated for millions of years by a single tree with short stubby leaves resembling those of a cypress, according to Olsen.

"It was probably a specialist in living under very hot conditions," he said.

But it wasn't just the plant world that suffered from this prehistoric climate change. Back at his fossil-strewn office at the Lamont-Doherty Earth Observatory, Olsen showed me the vestiges of the catastrophe in the animal realm. Although his work takes him as far as western China, the tristate area has some of the richest Triassic fossil sites in the world. He rifled through his extensive collection of local rocks to show me a series of ancient footprints he found with his twelve-year-old son along a Hudson River beach not far from his office. They demonstrated the striking reversal of evolutionary fortunes experienced by the ancient croc cousins and the dinos on either side of the extinction boundary. Before the mass extinction there were the oversized five-toed footprints of a vicious rauisuchian; this huge, athletic crocodile relative—built more like a tiger with scales than a croc—was the dominant predator of its time.

"So you can see it's almost three times the size of most of the dinosaurs that were around at the time," Olsen said.

After the mass extinction, these proportions reversed as the three-toed footprints of their understudy dinosaurs quickly assumed the massive proportions of popular imagination. They would remain that way for more than 135 million years. Meanwhile, only the meekest of the crocs made it through to the Jurassic—a line of runts that wriggled over the extinction boundary.

"Some of these croc relatives after the extinction are just absolutely adorable," Olsen said. "Honestly, they must have been really cute, almost doglike. But nothing we would recognize as crocodiles survived across the boundary. They had to reinvent that lifestyle in the Jurassic."

Though it wasn't nearly as extreme as the End-Permian, the End-Triassic mass extinction seems to have been a sort of Great Dying Jr., with huge injections of carbon into the atmosphere from volcanoes and a lethal super-greenhouse as the result. But the End-Triassic mass extinction might also serve as a gruesome template of sorts for our next few centuries.

"The timescales of these eruptions are appropriate comparisons for modern global warming and ocean acidification," Olsen said.

At the end of the Triassic, there's evidence for not only a heat wave on land but devastation in the seas as well. Bivalves (creatures like clams, scallops, and oysters) had largely replaced the brachiopods in the ocean after the Great Dying, marking an unglamorous but epochal transition for marine ecosystems. But half of the bivalves would go extinct at the end of the Triassic. Their shelled squid-like relatives, the ammonoids, once again almost completely vanish from the fossil record (in their typical fainting couch manner). And the flashy new ichthyosaurs were decimated as well.

But of the many sea creatures that wouldn't make it through the bottleneck of the End-Triassic mass extinction, the strangest might have been the legendarily enigmatic conodont. Conodonts are known primarily for their tiny, oddly baroque teeth, once described by *The New Yorker* writer John McPhee as "like wolf jaws, others like shark teeth, arrowheads, bits of serrated lizard spine—not unpleasing to the eye, with an asymmetrical, objet-trouvé appeal." The tiny fangs are interesting for two reasons: First, they're indispensable to oil companies. They change color when they're heated, illuminating "oil windows" in the rocks where conditions are perfect for generating petroleum. And second, for 150 years no one had any idea what the little trinkets were. Their ambiguity was enough to inspire science historian Simon Knell to write (without irony) that the identity of conodonts became to paleontologists "an object of mythology—an Arthurian sword in the stone by which all comers might test their intellectual strength." In recent reconstructions, the prickly knickknacks stuff the mouths of eel-like critters, fitting together with a ghoulish interlocking clockwork straight out of the Stan Winston creature workshop. Like the trilobites, conodonts had been true survivors, a wildly successful group that positively litter the fossil record for almost 300 million years, even surviving the Great Dying. And then, at the end of the Triassic, after eons of success, the conodonts suddenly vanished, leaving only their strange jaws behind.

"Conodonts are like God," German paleontologist Willi Ziegler once mused. "They are everywhere."

Until they weren't.

But the most striking feature of the End-Triassic extinction in the oceans was the wholesale destruction of corals.

"Coral reefs almost go completely extinct," Olsen said. "They just basically disappear from the planet entirely at the extinction."

University of Texas–Austin paleontologist Rowan Martindale's office is decorated with chunks of ancient reefs from around the world, including a block of Permian sponge hacked off the Guadalupe Mountains. Her work traces the fortunes of reefs throughout Earth history—a story of both stupendous successes and cataclysmic collapses. In the Triassic, it was both. Although they suffered losses during each of the Big Five extinctions, the reef collapse at the end of the Triassic was especially striking, coming as it did after one of the most spectacular reef-building episodes in Earth's history.

"In the latest Triassic, reefs do really well, and the classic case is the Austrian and German Alps," says Martindale. Martindale did her PhD work in these fairy-tale mountains, which are largely constructed from coral reefs that formed in the days when Europe huddled around the tropical Tethys Sea on the *east* coast of Pangaea. The hills surrounding Salzburg might be alive with the sound of music, but they're also dead with the eventual victims of the End-Triassic mass extinction.

"You hit the Triassic-Jurassic boundary, and for about 300,000 years there's no reefs and no corals in the rock record whatsoever," said Martindale.

Though it was 200 million years ago, the obliteration of reefs at the end of the Triassic is grimly resonant for the twenty-first century.

"The cool thing about the Triassic-Jurassic event is that it's the biggest hit ever to modern corals," Martindale said. "So that's why it's a big deal."

Earlier, more archaic reef systems in earth history—like the vast reefs of the Devonian, or the Permian limestones that loom over Texas—were relics from a different planet, strange patchworks of sponges, brachiopods, giant calcite horns, and honeycombs cemented together by calcifying algae. But the Triassic represents the birth of modern coral reef. Stony corals, of the sort that today make up reefs from Florida to Sydney, first appeared here in the Triassic—before nearly being swept clean from the fossil record forever.

Like a rerun of the End-Permian, what's particularly frightening is the culprit in this mass die-off: warmer, less oxygenated, and more acidic oceans responding in chemical lockstep to the huge injections of CO_2 then gushing from New Jersey and elsewhere.

"It's basically just this colossal collapse of reef systems," said Martindale. "If you live on a reef in the End-Triassic, chances are pretty good that you're going to go extinct."

Depending on how the next few decades go, the same might be said today.

"I was on the Caicos platform on the Turks and Caicos earlier this year, and we went to these reefs that were called 'the amazing reefs,'" she said. "They had just dredged a new channel for hotel boats and everything was dead. It was so bad."

To understand what happened in the oceans at the end of the Triassic, it's useful to look at modern coral reef systems, which have shrunk by perhaps 30 percent since the early 1980s (an appalling, geologically instantaneous lightning strike). Coral growth rates have slowed by 20 percent in the past two decades, and devastating bleaching events—what happens when warmer water forces corals to lose the microorganisms upon which they rely for food—have become common. Humans are currently increasing the carbon dioxide concentration in the atmosphere at a rate of 2 parts per million every year; if this trend continues and the oceans continue to acidify, coral reefs worldwide "will become rapidly eroding rubble banks" by midcentury, according to one landmark study. Reef diving after the End-Triassic mass extinction would bring a snorkeler back to this future, face-to-face with a world of broken slimy husks of coral that once billowed with Technicolor clouds of life.

As mentioned before, since the start of the Industrial Revolution, modern oceans have already reacted to atmospheric carbon dioxide by becoming 30 percent more acidic. Things like clamshells, the skeletons of corals and many types of plankton, and even the accelerometers in the heads of squid are made of calcium carbonate. You might be more familiar with calcium carbonate in its role as an antacid, or as chalk. You also might remember from elementary school science class what happens when you put a stick of chalk in acid. But not only does the ocean get more acidic when it's suffused with carbon dioxide, the altered chemistry also robs the ocean of carbonate by locking it up as biologically useless bicarbonate, making it unavailable to animals to build their shells and skeletons. Again, while politicians dither about the effects of excess carbon dioxide, all this remains fairly simple chemistry.

In more acidic, less carbonate-rich water, corals have a difficult time calcifying; they become less dense, more brittle, and more vulnerable to storm damage and predation; and they put more energy into making ever-weaker skeletons, siphoning away resources normally put toward reproduction. According to a 2007 study, researchers led by Ove Hoegh-Guldberg of the University of Queensland estimated that "reef erosion will exceed calcification at 450 to 500 [carbon dioxide] ppm." In other words, this is when the collapse of coral reefs and the animals that depend on them will begin in earnest. Given current carbon emissions trends, we are likely to reach this point by midcentury. Depressingly,

Hoegh-Guldberg and his associates used the lower end of the Intergovernmental Panel on Climate Change (IPCC) predictions for carbon dioxide emissions. In other words, the most optimistic scenarios yet seriously proposed in international climate negotiations will destroy the world's coral reefs, perhaps by midcentury. Hoegh-Guldberg noted that above 500 parts per million, corals stop growing altogether, and that more pessimistic emissions projections of 600 to 1,000 parts per million by the end of the century, as they put it suggestively, "defy consideration."*

Additionally, corals are exquisitely sensitive to temperature changes: many species cannot live in the cold, but they are also subject to life-threatening episodes of bleaching when the water gets too warm. Microorganisms called zooxanthellae live on reef-building corals (the corals first recruited these symbionts in the Triassic), and the corals rely on them to photosynthesize their food. When episodes of unusually warm water hit, zooxanthellae literally start poisoning this relationship and the corals, it's thought, expel them out of desperation. This is called "bleaching" for a good reason: to visit a reef after a bleaching event is to visit a panorama of calcium carbonate as white as desert bone. Bleaching is a medical emergency for corals, and the rare colony that does survive a bleaching event is often left an exhausted shade of its former splendor and even more vulnerable to future crises. Cores of centuries-old coral colonies show that the waves of global bleaching events that have wiped out corals in the past few decades—

* Though the coral reefs of the Triassic flourished under a much higher atmospheric CO_2 regime than present, Hoegh-Guldberg and his colleagues are quick to dismantle what could be used as a strawman argument by skeptics: "Although [modern] corals arose in the mid-Triassic and lived under much higher atmospheric CO_2 there is no evidence that they lived in waters with low-carbonate mineral saturation. . . . It is the rapid unbuffered increase in atmospheric CO_2 and not its absolute values that causes important associated changes such as reduced carbonate ions, pH and carbonate saturation of seawater."

like the horrific bleaching that pummeled the Florida Reef and the Hawaiian Islands in 2015—are unprecedented for at least the past several thousand years. Moreover, they only promise to intensify in the coming years. Rising sea levels could also effectively "drown" corals that are weakened by chronic stress and unable to move to higher ground, leaving the symbiotic organisms on which they depend for food unable to photosynthesize in deeper, darker waters. Combining this threat with the threats from overfishing and pollution, we can see why one ecologist has referred to reefs worldwide—which host 25 percent of species in the oceans—as a "zombie ecosystem."

"There's some pretty severe concern in the community about how reefs are going to look in fifty years," said Martindale. "People are talking about how we should just start freezing tissue."

Unfortunately, what will replace coral reefs is not nearly as photogenic. As University of Queensland biologist John Pandolfi put it, coral reefs worldwide are on a "slippery slope to slime," a panorama of lifeless, shattered mounds coated in green muck.

"There are already reefs that have started going over to fleshy algae," said Martindale.

"So the take-home is: depending on what we foresee for the future, in terms of how temperature changes or how pH changes, I mean we're really moving into a place where reefs are going to have a hard time sticking around."

Although, to date, most of the destruction of the coral reefs worldwide has been wrought by invasive species, pollution, and overfishing ("Florida's reefs are already relatively obliterated," said Martindale), the coming changes to ocean chemistry in the next century and the ensuing global reef collapse would be truly rare calamities in earth history. The terrifying reality of ocean acidification has only fully dawned on the scientific community in the last decade or so. Even more so than global warming, ocean acidification is what people who understand the fossil record, and who think about the future of the oceans, are most distressed by.

If another branch on the tree of life someday produces geologists, they might notice the sudden strange disappearance of coral reefs at our Pleistocene-Anthropocene boundary and compare it to the Triassic-Jurassic boundary 200 million years earlier. Depending on how clever these geologists of the far future are, they might even note the similarly wild swings in the carbon and oxygen isotopes in their rocks, pointing to a huge injection of carbon and a warming spike exactly at both extinctions. Though it might take only a few decades for us to wipe out coral reefs, if the End-Triassic mass extinction is any guide, these ecosystems will take not decades, centuries, or even millennia, but millions of years to restore. The decisions made in the next few years by the energy industry and the governments that regulate them will leave a record in the rocks that will last for hundreds of millions of years.

Considering how much we've already done to dismantle coral reefs, and projecting these trends forward into anything resembling geological time, it becomes clear why it's not unreasonable to compare what is going on today with the worst disasters in earth's history.

As we walked to Olsen's car in the parking lot of the Lamont-Doherty Earth Observatory, something started clanging just out of sight.

"I wonder if that's the beginning of our drilling project," Olsen said. "Oddly enough, that's related to attempts to sequester carbon. Oh, darn, we didn't talk about that."

Amazingly, the same cliffs under Olsen's office (the Palisades) that might have once caused the fourth major mass extinction in the history of complex life are now being recruited to stave off the sixth. If Olsen and Columbia colleagues Dennis Kent and Dave Goldberg are right, the same apocalyptic, carbon-belching basalt that destroyed the Triassic world could someday serve as a vast reservoir for anthropogenic CO_2—in an odd sort of penance for the cliffs and their ancient sins against the planet. The secret to this CO_2 burial would be a highly accelerated version of the weathering processes that have rescued the planet from extreme CO_2 greenhouses before.

"So the Palisades and the lava flows are also potential sinks for modern CO_2 because basalt reacts very rapidly with carbon dioxide to produce limestone," Olsen said. "So one way of sequestering CO_2 which may actually become a reality shortly is, you'd capture CO_2 before it enters the atmosphere at power plants and then you pump it into fractured basalt. Then it would convert to limestone fairly rapidly. So we've had some experiments here at [Columbia] to actually show that. We did an exploratory hole on exit 14 on the thruway last year, and now we're doing an exploratory hole here at [the Lamont-Doherty Earth Observatory] to determine if this is actually a possibility."

Such human ingenuity is one reason to be sanguine about our prospects for avoiding the cliff that every trend in geochemistry is currently pointing us toward. Undoubtedly, there are some obvious, unsettling parallels between the Triassic end-times and the current day—when, barring aggressive climate action, temperatures on the planet are expected to jump as much as 6 degrees, if not by the end of this century, then sometime during the next, with the oceans acidifying not on the scale of thousands of years but within decades. But even these terrible events in the geological record offer some reasons for hope. Stony corals, after all, survived the End-Triassic extinction; otherwise, they wouldn't be around today. In the same way, even if the worst projections come to pass, it's very unlikely that corals will go extinct altogether. The geological record is filled with plucky survivors that held out in remote areas (called refugia) where tolerable local conditions allowed them to wait it out until the worst was over. Perhaps some resourceful corals can adapt to extreme conditions and evolution will provide an off-ramp from extinction. If the geological record is any guide, it would take millions of years for the survivors to reestablish the large reef structures and ecosystems familiar to us today, but the planet is incredibly resilient. Take it from the plucky ammonites, who—though silent for millions of years in the fossil record after the End-Triassic volcanoes destroyed the planet—eventually timidly reappeared in the age of dinosaurs before exploding in a dizzying radiation of new shapes and sizes.

Another reason to take solace is that additional killers that might have been present at the end of the Triassic don't seem to threaten humanity in the short term. It's been suggested that the global warming of the End-Triassic destabilized huge stores of frozen methane at the bottom of the ocean that came bubbling to the surface. Methane is an extremely potent greenhouse gas, and when it degrades in the atmosphere, it becomes carbon dioxide. A catastrophic release of methane from the bottom of the ocean would have compounded what was already a climate catastrophe in the Triassic. Today similar reserves of frozen methane lurk in the cold dark corners of the ocean. University of Chicago geophysicist David Archer has written about the destructive potential of these deep-sea stores of carbon.

> If just 10% of the methane in hydrates were to reach the atmosphere within a few years it would be the equivalent of increasing the CO_2 concentration of the atmosphere by a factor of 10, an unimaginable climate shock. The methane hydrate reservoir has the potential to warm Earth's climate

to [extreme] hothouse conditions, within just a few years. The potential for devastation posed by the methane hydrate reservoir therefore seems comparable to the destructive potential from nuclear winter or from a comet or asteroid impact.

But if these methane hydrates were a menace in the Triassic, for now—despite their apocalyptic potential—methane hydrates in the modern seafloor seem to be fairly resistant to these kinds of catastrophic releases. And besides, the starting state of the Triassic was a *much* warmer planet than ours. It might have required a smaller push to tip the planet over into this sort of deadly feedback loop.

There are other reasons to think that the End-Triassic mass extinction might not be the best analogy for our modern challenges. Rutgers geochemist Morgan Schaller has calculated that flood basalt eruptions release enough sun-blocking sulfate aerosols to be equivalent to three Mount Pinatubo eruptions a day. Mount Pinatubo was the 1991 volcano in the Philippines that, when it exploded, lowered the global temperature by half a degree Celsius for three years. Pumping sulfate aerosols into the stratosphere is currently being pitched as a controversial geoengineering solution to global warming. (One reason it's controversial is that it does nothing to address ocean acidification.) In the End-Triassic, sulfates might have played a similarly brief but chilling role by temporarily countering the effect of all that carbon dioxide. Schaller contends that the result would have been volcanic winters in the tropical world. The sulfates would have lasted in the atmosphere for only a few years (which might explain why we don't find evidence for cooling in the fossil record), while the CO_2 super-greenhouse would then have kicked into high gear and lasted for thousands of years afterwards. In fact, if the planet did briefly cool, it would have suppressed weathering rates, allowing carbon dioxide to climb ever higher in these intervals. In this event, there would have been an even more extreme swing back to hot times when the sulfates eventually rained out of the atmosphere.

Given the prospects for brief blasts of cold during the mass extinction, Olsen has even proposed an explanation for the preferential survival of the dinosaurs—and extinction of the dominant crocs—that appeals to the increasingly plausible idea that perhaps all dinosaurs had feathers. This insulation, along with their unique physiology, would have allowed them to survive both the flash freezes and then the ensuing super-greenhouse. As yet, there's no evidence in the fossil record for brief volcanic winters—only the high-CO_2 hothouse lasting thousands of years. Nevertheless, it might have been, once again, a climatic whiplash between fire and ice that brought down the planet.

As the Triassic turned over into the Jurassic, after a million or so years of painful transition, life bloomed again. The dinosaurs colonized the niches relinquished by their departed rivals and eventually grew to become the majestic stewards of the planet in its most mythical age.

On the drive back to Boston from New York City, I passed a sign I'd seen many times along I-91 and this time could no longer resist. DINOSAUR STATE PARK, it read.

The unlikely landmark is just outside of Hartford, Connecticut, among the suburban subdivisions and office parks of the woodsy Connecticut River Valley. I pulled into Dinosaur State Park expecting to be underwhelmed and making wisecracks to myself about waspy, fiscally conservative Connecticut dinosaurs that played racquetball.

I stopped laughing when I walked into the park's geodesic dome near closing time and came upon the site's signature attraction: hundreds of dinosaur footsteps meandering about the sandstone floor, on the petrified shores of another rift valley lake. But this time, though still deep in the rifting heart of Pangaea, life on the planet was just on the other side of the mass extinction. This was the dawn of the Jurassic. The eruptions had quieted, and the recent apocalypse was evident only in the presence of this new roster of animals, who confidently governed the planet as if nothing had happened. The huge expanses of basalt had weathered away and drawn down carbon dioxide—as they always do, cooling the planet back down—and the lakes of lava that filled the rifting valleys of Pangaea had either become worn away or tucked into the vault of geology. The planet had pacified, and here in the Connecticut River Valley it had resumed its languid rhythms, only with a new ruling caste—the dinosaurs.

The footprints, at more than a foot long, were enormous compared to the dinosaur runts that came before the mass extinction. It's unknown who left these tracks (the same conditions that are good for preserving footprints aren't good for preserving dead bodies), but paleontologists suspect it might have been *Dilophosaurus*, a huge dinosaur more than 20 feet long (one that the movie *Jurassic Park* inexplicably turns into a dog-sized frilled lizard that spits poison phlegm). Although these gigantic three-toed footprints were strewn all across this lakeshore, the prints of the killer crocs of the Triassic were nowhere to be found.

I was alone with the huge dinosaur footprints, which were lit in sharp relief from a low angle, as unseen speakers pumped in the evocative sounds of primal humidity: the hum of droning insects, distant rumbles of thunder. A tropical, cycad-lined lakefront mural framed the trackways and the model figures of two 20-foot dilophosaurs stalked the exhibit, their enormous feet pressing into the wet sand, as they surveyed their erstwhile haunts with purpose.

I found myself almost embarrassed by how deeply the pockmarked slabs moved me. There's something about fossil footprints that are strangely personal, perhaps even more so than the bones themselves that animals offer up to the ages. Unlike the plaster museum reconstructions of dinosaurs, which are often contorted in poses of theatrical menace, these footprints were utterly undramatic and prosaic. There was no pretense in these footfalls. This animal was utterly unaware of its place in the history of life. This was not a tableau of life in the Jurassic, but of life on a Tuesday afternoon. Here the footprints stop. There they resume in another direction. Here they break into a widely spaced jog, and there they narrow to a halt. These were actual moments of indecision recorded here in the rock—whims and lost trains of thoughts in the skulls of these unspeakably ancient animals as they prowled the shore. These were individuals, it struck me, each with its own personality and biography. I was unexpectedly encountering these personalities here, if only for a few moments—moments that the creatures themselves were blithely unaware would be preserved for all time. It was enough to make me forget about the unbridgeable chasm in time and space that separated us—until I heard the muffled cry of a car alarm going off in the parking lot.

Next to me a woman and her boyfriend approached the exhibit with the same unexpected reverence. Her bedazzled iPhone and Insane Clown Posse T-shirt weren't suggestive (though I could be wrong) of a lifetime spent pulling holotypes from museum collections. But the humbling reproach of deep time here was intoxicating.

"What is there going to be left of us?" she asked her energy drink–chugging boyfriend. He put down his can and looked at her searchingly.

"What will we leave behind?" she pressed.

6

THE END-CRETACEOUS MASS EXTINCTION

66 Million Years Ago

Should a comet in its course strike the Earth, it might instantly beat it to pieces. . . . But our comfort is, the same great Power that made the Universe, governs it by his providence. And such terrible catastrophes will not happen till 'tis best they should. In the mean time, we must not presume too much on our own Importance. There are an infinite Number of Worlds under the divine Government, and if this was annihilated it would scarce be miss'd in the Universe.

—Benjamin Franklin, 1757

It seems almost unfair to dwell now on the demise of the dinosaurs rather than celebrate their spectacular reign. Dinosaurs flourished, adapted, diversified, dominated, and, most impressively, *endured* for an incomprehensible length of time. Modern humans have been on this planet for far less than a million years. Dinosaurs for more than 200 million. This epic time span can make for some head-scratching chronology: *T. rex*, the iconic superpredator of the Cretaceous, lived far closer in time to human beings than it ever did to the Jurassic showstopper *Stegosaurus*.* Even the supposed downfall of the dinosaurs is not all it seems: modern birds are both indisputably dinosaurs (theropods, just like *T. rex*) and vastly more species-rich than mammals.

"There are twice as many species of birds as there are mammals," Paul Olsen told me. "So we're still living in the age of dinosaurs. Mammals have never been as successful as dinosaurs. Still aren't."

Some might see humanity as the end member of an inevitable progression to more advanced life. But this comforting view doesn't square with the brute fact of 136 million years of meek mammal serfdom in the shadows of dinosaurs—an arrangement that required an inconceivable catastrophe to upend.

As Walter Alvarez writes, "[The Mesozoic] was a stable world. There is every reason to believe that if it had remained undisturbed, [it] could have continued indefinitely, with the slightly evolved descendants of the dinosaurs dominating a world in which humans never appeared."

Dinosaurs are the protagonists so far in the history of animal life on land—not some peculiar preamble to our own story. Throughout the epochs they inhabited every niche—predator and prey, herbivore and carnivore—and spanned every size,† from the pigeonlike anchiornis to the hangar-sized *argentinosaurus*. Sauropods like these were so monumental that their methane farts might have been partly responsible for making the Mesozoic so warm.

Dinosaurs herded along tropical beaches under a withering sun and darted through lush polar forests under the spectral glow of the aurora borealis—and in Antarctica under the aurora australis.

* Notwithstanding science-fair dioramas depicting battles between the two.
† The estimated masses of known dinosaurs range over six orders of magnitude.

It is this utter dominance that makes their fall at the end of the Cretaceous one of the most intensively studied and mythic events in planetary history. In fittingly sensational form, their deathblow was catastrophically abrupt and mind-bendingly spectacular.

At the end of the Cretaceous, the largest asteroid known to have hit any planet in the solar system in a half-billion years hit Earth . . .

At virtually the same time that one of the largest volcanic eruptions ever smothered parts of India in lava more than *2 miles deep*.

"It's not like dinosaurs were the only thing that went extinct," said Tom Williamson, curator of paleontology for the New Mexico Museum of Natural History and Science. Williamson and I fumbled with camp stove–fired fajitas and Tecates after a long day of fossil collecting in the northwest New Mexican desert. The desert is Williamson's second home, and he was joined here this summer by an NSF-funded crack squad that included scientists from the Universities of Nebraska and Edinburgh and Baylor University. The team—a modern mix of geochemists, paleontologists, magnetostratigraphers, and geochronologists—sought to uncover in these rocks how a shattered world reassembled itself in the immediate aftermath of the most famous mass extinction in earth history.

"Tons of mammals went out," Williamson said, looking out over the candy-striped cliffs and canyons of Angel Peak Scenic Area. "It almost wiped out the marsupials. Tons of birds died."

Williamson's point—that the extinction of the dinosaurs was only part of the story of the End-Cretaceous mass extinction—was well taken. The week before, I had joined University of Alabama paleontologist Dana Ehret to comb through marine chalk just outside of Selma, Alabama, in search of bones from unspeakable 60-foot sea monsters called mosasaurs, which ruled the oceans of the Cretaceous. Mosasaurs all died too. The ferocious reptiles swam alongside, and occasionally devoured, giant ammonoids, whose tentacles and stately spiral shells had haunted the oceans for hundreds of millions of years since the Devonian. Ammonoids all died too. On the seabed were giant, bucketlike, and boomerang-shaped clams—the rudists—that created huge reefs visible today in the white cliffs of southern France and in thick seams that run for miles through the Pyrenees. Rudists all died too. Farther offshore, pencil-necked plesiosaurs paddled by, while above the waves, giraffe-sized pterosaurs with wingspans as broad as airplanes glided by overhead—mocking the best efforts of biomechanics modelers. All these creatures, in and above the sea, represented the ancient earth at its most outlandish, fever-dream best. And they all died in a geological moment.

And on land, so did the dinosaurs and almost everything else alive.

As the sun set over New Mexico it lit the desert badlands in a melancholy twilight blush so arresting it merited comment.

"Isn't this unbelievable?" Williamson said, gazing out at the canyons. "Anywhere else this would be a national monument. Here it's an oil field. It's like the Grand Canyon of New Mexico and no one even knows about it."

In the badlands below, the ragged access roads cutting through the landscape led inevitably to oil pump jacks, which sucked ancient sunlight from the ground.* In the distance, a thin wisp of opaque, yellowish smog traced a path across the horizon.

"That's the smog layer from the Four Corners coal-burning

---

* In 2014, NASA discovered a cloud of methane hovering over this corner of the state, steadily leaking—not from brontosaurs, but from the region's coal-bed methane industry at an annual rate of 600,000 metric tons.

power plant," Williamson said. "They're burning Cretaceous coal from New Mexico—trees the dinosaurs were eating."

Though smoke from the dinosaurs' trees hung in the air above, there were no more dinosaurs in the rocks below. The badlands—banded by grays, purples, tans, blacks, and reds—resemble those farther south in the San Juan Basin that are filled with the myth-making femurs of tyrannosaurs and titanosaurs. But these canyons are stocked with more modest fossils from just barely after the End-Cretaceous mass extinction, in the shocking hangover after the age of reptiles. The grand prize in these hills is not a *T. rex* skull the size of a bumper car, but tiny teeth from weasel-like survivors. Squinting at the dusty badlands, I tried to imagine the lazy streams, oxbow lakes, forests, and marshes where timid mammals grew ever larger and more confident in claiming their new world.

After sundown, the sunburned crew cracked jokes and argued about sports, fueled by the easy intimacy of a crackling campfire. University of Edinburgh paleontologist Steve Brusatte, an Illinois native living in the UK who struggles to catch his beloved Bulls and Blackhawks games abroad, was an eager participant. But eventually the conversation returned to the animals buried in the desert.

Regrettably, most of us lose interest in dinosaurs with age—even after feverish childhood obsessions—but for Brusatte that enthusiasm has never waned. His recent research focuses on the rise of the tyrannosaurs, a group that for most of their 100-million-year history were human-sized and marginalized as other, more primitive groups, like allosauroids, enjoyed their perch at the top of the food chain. But something very bad might have happened earlier in the Cretaceous period to clear the tyrannosaurs' path to the top, almost 20 million years before the iconic mass extinction. Under the ocean, huge pulses of lava issued from the Caribbean, from

Madagascar as it tore away from India, and from the largest flood basalt in the world, the Oontong-Java Plateau, a monster volcanic province that burbled deep in the heart of the Pacific. The eruptions once again seem to have rendered huge swaths of the ocean anoxic and legions of sea life extinct and, on land, might have even driven climate change that toppled the allosauroids. Whatever happened to them, the runt tyrannosaurs of North America and Asia assumed the throne in their wake—and quickly became the biggest, baddest things to ever walk the earth.

Though he's not exactly a disinterested party, I asked Brusatte if *T. rex* deserves its awesome reputation.

"I mean, *T. rex* is, as far as we know, the biggest predator that's ever lived on land," he said. "Today it's a polar bear. *T. rex* could stomp on a polar bear."

"There were other dinosaurs that kind of got up to *T. rex*'s size as predators, but none of them were quite as big and bulky. It really is that icon, and it deserves its place—I mean, it's 13 meters long and 7 tons," he said, laughing at the absurd dimensions. "There's nothing alive like that today."

"It would have had binocular vision like we do," Brusatte continued. "It had big optic lobes in its brain. It had huge olfactory lobes, so it could smell really, really well. It had an inner ear that could hear low-frequency sounds. It was an intelligent animal. It had a pretty big brain. It wasn't just a brawny thing. It was a pretty brainy thing too."

I imagined that when a *T. rex* attacked it did so, not with the dopey, glassy-eyed, perfunctory gaze of a shark, but with a frigid, purposeful menace—with the righteous determination of a gigantic bird that wants to kill you. For Brusatte, though, the most interesting thing about *T. rex* isn't its turbocharged biology, but its meteoric rise at the end of the Cretaceous, and its even more precipitous (and rather more literal) meteoric fall.

"One thing about tyrannosaurs that a lot of people don't necessarily realize is that *T. rex* really was the last of the dinosaurs," he said wistfully. "It was there when the asteroid hit. Something as dominant and iconic as *T. rex,* it goes out right away. Or even if it was the volcanoes that did it, it's still incredibly sudden. You have this great dinosaur, and it just goes away, and within a few tens of thousands of years you have an incredible diversity of new mammals. Nothing near the size of *T. rex,* but in geological terms that is a *knife edge*. You go from a world dominated by these big dinosaurs, with *T. rex* at the top of the food chain, to what we see here. If you think *T. rex* is the ultimate dinosaur, the ultimate predator, that wasn't enough to get it through what happened at the end of the Cretaceous."

So what happened at the end of the Cretaceous?

The impact of the paper "Extraterrestrial Cause for the Cretaceous-Tertiary Extinction" on the science community would be difficult to exaggerate. Before 1980, the death of the dinosaurs was shrouded in more or less abject ignorance—an ignorance betrayed by the spread of truly insane theories about their demise. Alan Charig, a curator at the Natural History Museum of London, once compiled eighty-nine proposed culprits he had heard suggested during his tenure. They included:

"Disease, nutritional problems, parasites, internecine fighting, imbalance of hormonal and endocrine systems, slipped vertebral disks, racial senility, mammals preying on dinosaur eggs, temperature-induced changes in the sex ratios of embryos, the small size of dinosaur brains (and consequent stupidity), and suicidal psychoses."

Other killers, proposed with varying degrees of seriousness, included death by AIDS from outer space and an epidemic of terminal constipation from ingesting the recently flourishing flowering plants.*

In 1980, geologist Walter Alvarez and his Nobel Prize–winning physicist father Luis blew a hole in the scientific community (and this raft of inane speculation) with a discovery that upended 150 years of geology.† Reviving the moribund spirit of catastrophism, the Alvarezes discovered evidence in the rock record that pointed to biblical destruction at the end of the age of dinosaurs.

Working in the scenic Apennine mountains outside of the postcard-perfect medieval Italian town of Gubbio, Walter Alvarez puzzled over the sudden, nearly total extinction of plankton in between Cretaceous and Tertiary rocks in an outcrop of limestone thrust up from the bottom of the ocean. The layers were separated by a curiously fossil-free layer of clay, and Alvarez wanted to know how long this interval, which seemingly upended life on earth, lasted. This curious break in the rocks is what's known in geology as the Cretaceous-Paleogene (K-Pg)—or to use a more dated but still widely used term, the Cretaceous-Tertiary (K-T) boundary.‡

In geology, the thickness of a rock layer is often a misleading indication of how quickly it was deposited, but there was no reason to doubt that this transformative gap might have spanned ages. The famous early geologist Charles Lyell himself had reasoned as much more than a century earlier when he noted the profound

---

* The satirical newspaper *The Onion* recently joined in on the fun, dutifully reporting that "Paleontologists Determine Dinosaurs Were Killed by Someone They Trusted."

† Even the Alvarez paper points to the proliferation of bizarre theories in circulation at the time, mentioning one that relied on "the flooding of the ocean surface by fresh water from a postulated arctic lake."

‡ The K in K-T and K-Pg stands for *Kreide,* the German word for "chalk." The letter C could not be used as an abbreviation for Cretaceous because it was already in use for the Cambrian period.

break in life between the Cretaceous and the Tertiary layers but explained away the disruption with what was obviously millions of years of missing rock.

To solve the riddle once and for all, Alvarez and his father devised an ingenious method to determine how much time had passed in the barren clay layer. Although he never suspected that a killer asteroid might be involved, the elder Alvarez reasoned that dust from harmless meteor showers would fall at a tiny but constant rate over the earth over the course of millions of years. If they measured the amount of the trace element iridium—a component of this dust—in the layer, one of two things would be apparent. If they found no iridium whatsoever, they would know that whatever happened between the Cretaceous and Tertiary happened too quickly to allow the steady rain of this extraterrestrial dust to accumulate in the disaster layer as it was deposited. Conversely, a small accumulation of the rare metal would mean that a great expanse of time had passed and that the changes at the end of the Cretaceous had happened gradually. They sent their Italian samples to the Lawrence Berkeley National Laboratory to be analyzed by ace nuclear chemist Frank Asaro and his handy nuclear reactor, and awaited the results.

What they found made no sense. Yes, there was iridium in the samples, but there was almost 100 times more than they expected. The most plausible explanation, then, was not a light rain of space dust over the ages, but one sudden, catastrophic pummeling from the heavens.

At the same time that the Alvarezes were investigating Gubbio, Dutch paleontologist Jan Smit—similarly curious about the sudden change in plankton at the K-T in limestone in Caravaca, Spain—independently discovered the iridium layer. The Alvarezes published first and were immortalized in one of the most cited papers in the history of geology. Jan Smit doesn't have a Wikipedia page.*

Reaction to the idea that the dinosaurs were killed by a space rock ranged from justified scientific skepticism to puzzlingly uninformed proclamations, like that handed down by the *New York Times* editorial board, which scoffed: "Astronomers should leave to astrologers the task of seeking the cause of earthly events in the stars." To which Walter Alvarez replied, in a letter to the editor: "May we suggest it might be best if editors left to scientists the task of adjudicating scientific questions?"

Still, the iridium layer alone was not enough to convince everyone, and much of the 1980s was spent in heated, often bitter dispute, especially between paleontologists weaned on Lyell's uniformitarianism and what they saw as an upstart band of physicists and astronomers who presumed to explain their beloved fossil record back to them. Many of these old antagonisms linger even to this day: one geologist I interviewed agreed to help answer questions on all of the mass extinctions except for the K-T, calling the extinction "too political." Though Walter Alvarez blames much of the acrimony in the debate on a yellow press corps hungry for scandal, his camp was responsible for its share of infelicitous quotes, including a number of hilarious broadsides from his father.

"I'd say he's a weak sister," Luis Alvarez told the *New York Times* about an academic rival who promoted volcanism as an alternative explanation for the extinction. "I thought he'd been knocked out of the ball game and had just disappeared, because nobody invites him to conferences anymore." In the same interview, the elder Alvarez famously scoffed, "I don't like to say bad things about paleontologists, but they're really not very good scientists. They're more like stamp collectors."

---

* Alvarez graciously calls Smit the "co-discoverer" of the iridium layer.

"Opponents in this debate," one scientist lamented, "have been reduced to name-calling."

For eleven years after the publication of "Extraterrestrial Cause for the Cretaceous-Tertiary Extinction," the chorus of asteroid skeptics asked "Where's the crater?" while impact proponents scoured the globe for impact structures. Even the discovery of shocked quartz grains at the K-T boundary, which could only have been created by a violent impact (or nuclear weapons testing by dinosaurs), did little to quiet the doubters. There was the disconcerting possibility that the crater might never be found. Perhaps the asteroid had landed in the ocean and the crater it punctured in the crust had since been chewed up in a subduction zone along the edge of the earth's tectonic plates, where ocean crust is continually being thrust back down into the furnace to be recycled.

Then hints slowly began to trickle in from the field that researchers were closing in on the crater.

I thumbed through the fine print of old geology papers looking for map coordinates before contacting a longhorn ranch owner outside of Waco, Texas. I asked whether I could poke around his property to look for tsunami debris from the asteroid that killed the dinosaurs. Amazingly, he said yes. Here in the middle of Texas, geologists had found a jumble of strange rocks that pointed the way to the apocalypse, and I wanted to see it for myself.

The land in southeast Texas is wide and flat and grades forward through geologic time as you head toward the coast. The Brazos River, which skirts by Houston on its way to the sea, serves as a sort of River Styx through earth history, and a kayak trip down it provides opportunities to hop out and sample fossils along its banks in chronological order. Under a highway bridge spanning the river, I collected 50-million-year-old seashells and shark teeth. If I had ventured closer to the coast, I would have found the remains of more recent creatures, like mammoth teeth, or the bony plates of supersized armadillos exposed on point bars under the Texan sun. But I was headed upstream and back in time.

I wasn't far from the Creation Evidence Museum in Glen Rose, Texas, a 501(c)(3) outfit led by a director whose website bio hinted at a lifetime spent pursuing endearingly wacky ideas in an attempt to reconcile the language of modern science with Iron Age origin myths—like his slick-sounding "Crystalline Canopy Theory," which proposed to provide a technical account of how the Firmament was made on Day Two of Creation, 6,000 years ago. While the Texas oil economy relies on the truth of geology, many of its inhabitants remain stubbornly resistant to its charms.

When I arrived at the electrified gate of the ranch, I met ranch owner Ronnie Mullinax, a man of few words and a mascot of Texas independence, wearing a cowboy hat, boots, denim, wraparound sunglasses, and, on his waist, the largest handgun I'd ever seen. We didn't share much in the way of pleasantries, but he was extremely generous with his time, given that I'd invited myself onto his land with a strange request. He graciously agreed to drive me and two scientists I had recruited from Texas A&M on his ATV to the K-T site. We bumped along down a dirt path, over streams and fields—pausing to marvel at his prized steers—to the woods at the far edge of the property, where we stopped. He told us to get out and promptly took the long-barreled gun out of the holster.

"For snakes," he reassured me. He led us into the woods and down into a gully where a peaceful brook tumbled over a small outcrop of some of the weirdest rock I'd ever seen. Here, midway between College Station and Waco, over a small waterfall in the woods, the Mesozoic abruptly became the Cenozoic. I had recruited a paleontologist and geologist to help me understand what I was looking at, but they were as mystified as I was by the inscrutable chaos in the rocks.

It was once again Jan Smit who first proposed that this jumble was the fallout from killer waves that sloshed around the Gulf of Mexico following an enormous asteroid impact somewhere nearby. At the bottom of the little waterfall in the woods was a rowdy layer of broken blocks of limestone, cemented together in a wild jambalaya. This, Smit said, was the initial devastation of the tsunami as it tore up the seafloor and chunks of the earth itself, ejected by the impact, came crashing down from above. Above this confusion of rock was a thick layer of sandstone, laid down as the ocean was still sloshing with sand scoured from beaches and loosed from landslides before it settled to the bottom in the hours and days after the tsunamis. At the top of this sandstone was a pencil-thin layer that glittered like gold, where finer particles could settle out of the water only after the drama subsided and the seas calmed once again.

Working on rocks in Beloc, Haiti, that originally formed at the bottom of the ocean, Haitian-American geologist Florentin Maurasse found a similarly strange bed of sand dumped at the K-T boundary. Other tsunami sections were soon discovered in Cuba and northeast Mexico. Clearly something very bad had happened somewhere in the Gulf of Mexico. K-T researchers were getting tantalizingly close to their impact.

Geologist Mario Rebolledo of the Centro de Investigación Científica de Yucatán (CICY) is intimately familiar with the gigantic crater that nearly wiped out life on earth at the end of the Cretaceous. Not only does he study the 66-million-year-old structure for a living, he also lives inside it. Rebolledo met me in Mérida, the sprawling capital of the Mexican state of Yucatán. The city of a million radiates from a Spanish colonial center, complete with charming pastel mansions, cobblestone streets, and cathedrals—

and sits well inside the 110-mile-wide impact crater that ended the age of reptiles. The crater—eroded and buried by millions of years of marine limestone—isn't visible at the surface; instead, it forms a complex, concentric bruise in the basement of the Yucatán Peninsula, reaching far into the Gulf of Mexico. It's the largest crater known on earth from the last billion years, and it was made in the same geological instant that the dinosaurs, the gigantic swimming reptiles, the pterosaurs, the ammonites, and much else living on the planet were exterminated. We talked dinosaur end-times over mole poblano.

Rebolledo recounted the peculiar story of the crater's discovery, which hid in plain sight for decades. In 1950, oil-hunting geophysicists working for the state-owned Mexican petroleum company Pemex discovered an enormous circular structure underneath the Yucatán Peninsula. Upon pulling up melted rock from a drill core, they dismissed it as ancient lava and reasoned that the whole structure was some sort of gigantic buried volcano. As volcanoes aren't the best place to prospect for oil, the curious structure was ignored for decades.

After the Alvarezes published their seminal paper in 1980, a frantic global search for the crater ensued. The search lasted more than a decade, though it needn't have lasted even a year. In the late 1970s, geophysicists Antonio Camargo and Glen Penfield, working again for Pemex, reevaluated the structure under the Yucatán with gravity surveys of the region and realized that it looked nothing like a buried volcano. While the K-T community was at a conference in Snowbird, Utah, in 1981 wondering where the crater was, Penfield and Camargo presented a paper at an oil industry meeting in Houston, asserting that the craterlike structure radiating from the coastal Mexican town of Chicxulub was in fact a crater, and most likely the same one that killed the dinosaurs.

Carlos Byars, a reporter for the *Houston Chronicle* who was in the audience, wrote up a story documenting the find—one that continued to utterly escape the notice of paleontologists for a decade. Ten years later, Byars once again found himself in the audience at a geology conference when geoscientists mused aloud about the whereabouts of an astounding crater that—given the evidence for tsunamis in the Brazos River and elsewhere—must exist somewhere in the Gulf of Mexico.

"Byars literally stood up in the audience and said, 'I know where it is!'" said Rebolledo. "Everyone looked at him like 'Oh, this guy's crazy.' So he called his office and had his story from ten years ago faxed over to show to the people in the room."

After a decade of searching, the impact crowd finally had their crater.

"I don't think Carlos got the credit he deserved," said Rebolledo. "At some point someone has to give him more credit."

I told Rebolledo that, as a fellow journalist, I'd be happy to oblige. But what sort of catastrophe did these newfound scars in the rock imply?

"The meteorite itself was so massive that it didn't notice any atmosphere whatsoever," said Rebolledo. "It was traveling 20 to 40 kilometers per second, 10 kilometers—probably 14 kilometers—wide, pushing the atmosphere and building such incredible pressure that the ocean in front of it just went away."

These numbers are precise without usefully conveying the scale of the calamity. What they mean is that a rock larger than Mount Everest hit planet Earth traveling twenty times faster than a bullet. This is so fast that it would have traversed the distance from the cruising altitude of a 747 to the ground in 0.3 seconds. The asteroid itself was so large that, even at the moment of impact, the top of it might have still towered more than a mile *above* the cruising altitude of a 747. In its nearly instantaneous descent, it compressed the air below it so violently that it briefly became several times hotter than the surface of the sun.

"The pressure of the atmosphere in front of the asteroid started excavating the crater before it even got there," Rebolledo said. "Then, when the meteorite touched ground zero, it was totally intact. It was so massive that the atmosphere didn't even make a scratch on it."

Unlike the typical Hollywood CGI depictions of asteroid impacts, where an extraterrestrial charcoal briquette gently smolders across the sky, in the Yucatán it would have been a pleasant day one second and the world was already over by the next. As the asteroid collided with the earth, in the sky above it where there should have been air, the rock had punched a hole of outer space vacuum in the atmosphere. As the heavens rushed in to close this hole, enormous volumes of earth were expelled into orbit and beyond—all within a second or two of impact.

"So there's probably little bits of dinosaur bone up on the moon?" I asked.

"Yeah, probably."

Along with researchers from the University of Texas at Austin and Imperial College London, Rebolledo is part of a $10 million expedition to drill thousands of feet down, past the quiet limestone snowdrift of the Cenozoic and into the maelstrom of rock created by this cataclysm. In particular, Rebolledo's team will drill into the so-called peak rings in the interior of the crater. These peak rings—essentially rings within rings in the interior of the crater—do not occur in your run-of-the-mill impact.

Craters might seem like fairly straightforward phenomena—namely, large tidy bowls punched into the earth by stray fastballs from outer space. But when the earth isn't merely pocked by errant pebbles but demolished by miniature worlds, the scars left behind take on more interesting forms. Truly enormous craters—like the one in Mexico and a few other rare spots in the solar system—transform whole landscapes of frangible rock into almost a liquid, and the countryside rolls and bounces like Harold Edgerton's strobe-lit milk drops. In Chicxulub the asteroid instantaneously put a hole in the ground more than 20 miles deep—deep enough, astoundingly, to puncture the earth's mantle—and stretching more than 60 miles wide. Over the next few unimaginable seconds, the earth behaved like the surface of a pond after a rock has been thrown in: complex peaks and ripples resonated throughout the Yucatán before being frozen in place as crazy, ready-made mountain ranges that would have loomed over the crater floor as high as the Himalayas.

Rebolledo and company hope to shed light on both the exotic physics and geology of the impact and the unlikely recovery of life in the traumatized landscape. Some intriguing hints about this fire-and-brimstone post-impact world have been teased out of another drill core in the area, carried out in nearby Yaxcopoil in the early 2000s. There, under thousands of feet of limestone, was the sudden confusion of shattered rock from the impact itself. But amid this jumble was also a strange mélange of minerals more familiar to deep-sea hydrothermal vent systems. Hydrothermal vents bubble up near underwater volcanoes and midocean ridges where seawater is able to mix with the infernal, convecting world beneath the eggshell of earth's crust. In the dinosaur crater, this roiling world began when the immediate chaos of the impact subsided and the ocean waters roared back into the newly made Mexican Hades. As mammals on land inherited a ghost world of vanished giants, this enormous gash in the Yucatán remained hot for 2 million years after the extinction, as a boiling headstone for the Mesozoic.

When a team from the Woods Hole Oceanographic Institution

first visited a hydrothermal vent at the bottom of the ocean off the Galápagos in 1977, they were dumbfounded to find an entire ecosystem—from bleached-white crabs to tube worms—far from the life-giving bath of sunlight, supported instead by chemosynthetic bacteria feeding off the metal-rich brew spewing out of the earth. Since that groundbreaking expedition, hydrothermal vents have been posited as candidate locations for the origins of life on earth. Rebolledo's drilling expedition will search for the fossils of extreme microbes living in the disaster zone immediately after the impact, motivated by the thrilling idea that if Chicxulub did in fact host this hardy life, perhaps similar craters did the same billions of years ago for early life on earth. Four billion years ago, as the solar system was still sorting itself out, impacts like the one that struck the Yucatán were humdrum affairs—the largest impactors measured, not in miles, but in comparisons to other planets. The ocean of the early Earth was deeply inhospitable, but perhaps the hydrothermal niches carved by gigantic asteroids were not. Perhaps, rather than crime scenes implicated in mass death, enormous-impact craters were the cradles for life in its infancy.

Still, I wanted to know more about the effects of the impact itself. What was it about the collision that actually killed the dinosaurs? As gruesome as the carnage was in Mexico, a 110-mile hole in the ground doesn't explain why the remaining 170 *million* square miles of the planet were virtually sterilized. I called up one of the world's preeminent impact modelers, Jay Melosh of Purdue University. For Melosh, the connection is clear.

"Basically every species on earth and certainly almost every animal died," he said. "And I think it was probably on the day of the impact."

In this version, there isn't anything subtle or complicated about the manner of the dinosaurs' death.

"Most dinosaurs were literally roasted in their tracks," he said.

One of the first, seemingly sensible questions to ask about the impact is: what did it look like? But it's an almost meaningless question. If you could see the impact, you were dead.

"What you'd see first, if you were within a few thousand kilometers of the impact, is the fireball," said Melosh. "And the first thing you'd do is be blinded and everything around you would be set on fire."

When a rocky envoy from outer space visited Chelyabinsk, Russia, in 2013, shattering windows and spawning dozens of dash-cam YouTube clips, the damage the meteorite caused took many by surprise.

"Even with Chelyabinsk, the people who were looking at the fireball were temporarily blinded," Melosh said. "It also emitted a lot of ultraviolet, and people got sunburned just from the UV. And this was for a tiny 20-meter object that dissipated its energy in the atmosphere."

Chelyabinsk released an amount of energy equivalent to half a megaton of TNT. Chicxulub released 100 million megatons.

"There's no real way to internalize that number," Melosh said. "It's certainly enough to lift a mountain back into space at escape velocity."

For the dinosaurs of coastal Alabama, the show would have been over rather quickly: as soon as the strange soundless fireball appeared on the horizon they would have been dead. But for those far enough away to be shielded from this lethal curtain of light, news of the impact would reach them soon enough.

"So the first thing is the radiation from the fireball, which is so hot, it's mostly optical, but then you get the ejecta arriving."

A phenomenal amount of earth was excavated from the crater. This was the ejecta, so named for its literal ejection into orbit.

Loosed from the surly bonds of earth for a moment, the rock followed intercontinental ballistic trajectories to the far reaches of the globe. When it returned, it burned up in the atmosphere in a worldwide blizzard of meteorites. This is one of the mechanisms that provides the asteroid theory its globally lethal thrust.

"It covered the earth within about an hour," said Melosh. "As it started falling, the sky would have turned red, and it would have gotten oppressively hot. And then it would have gotten hotter, and hotter, and hotter."

Melosh and his colleagues calculated that the incoming rock would have subjected the surface of the earth to 10 kilowatts per square meter of energy. He then turned to his appliances to figure out what, exactly, that meant.

"What I did was, I measured the energy input to my oven on different settings, and I could get about 7 kilowatts per square meter at 'broil,'" he said. "So that gives you a feeling of what it would have been like."

The broiling would have lasted 20 minutes.

"Any animals that couldn't seek shelter would have been literally roasted," he said. "That explains a lot of the survival."

Melosh was originally concerned that the survival of some lineages, like birds, falsified the theory. Modern birds spend their time out in the open, where they would have been incinerated, unlike the surviving mammals that could have burrowed to avoid the inferno.

"But it turns out, all modern birds are descended from an order of water birds, whose modern relatives nest in holes in the bank," he said. "So they were probably able to survive that way. They had burrows that they could hide in. The impact happened in June to July, so they were probably nesting at the time."

Wait—what?

When geologists can nail something down to within a few

hundred thousand years, it's considered a triumph of fine-scale geochronology. How could we possibly know what *month* the asteroid struck? Melosh points to the work of paleobotanist Jack Wolfe, who, in studying water lily and lotus remains at the K-T boundary in Teapot Dome, Wyoming, purported to find the seeds of the water lilies in the section, but none from the later-blooming lotuses, placing the impact sometime in early June.

"Then you'd get the seismic shaking," said Melosh.

"It would be comparable to a magnitude 12 earthquake. Which . . . well, there's no such thing as a magnitude 12 seismic earthquake because the elastic strain [of the earth's crust] can't contain that much energy, but it is certainly possible for a big impact."

Tremendous ancient landslides discovered by oceanographers, like the slumped deposits found at the edge of the continental shelf in locations like Blake Nose off the Carolinas, date to the end of the Cretaceous. As a team of paleoceanographers from Woods Hole, Texas A&M, and the University of Edinburgh put it:

"Much of the eastern seaboard of North America must have catastrophically failed during the [K-T] impact event, creating one of the largest submarine landslides on the face of the Earth."

The earthquakes would have been quite noticeable even on the opposite side of the planet. As a geophysicist later put it to me, a magnitude 11 to 12 earthquake at any one location would feel like a magnitude 9 earthquake everywhere else on the planet.

"And then finally there was the air blast," said Melosh.

In 1908, when a 200-foot space rock hit in the middle of nowhere Siberia (thankfully), the air blast from the explosion flattened entire forests for hundreds of square miles. Melosh is no stranger to air blasts, having experienced one at relatively close range at the invitation of the US Army outside of Lake Havasu City, Nevada, where the army was studying the effects of shock waves. Melosh had the opportunity to witness 500 tons of high explosives detonate from a kilometer away.

"It was quite impressive, you can actually see the shock wave in the air," he said. "It looks like a shining bubble that expands in complete silence. It expanded very, very rapidly until it reached us; then you could hear the *kaboom!* But before you heard that, you felt the shaking in your feet because the seismic energy propagates faster than the sound in air. So you feel the shaking in your feet, you see this shimmering bubble—sort of like a soap bubble—expanding really fast, and then you hear the *kaboom*. We were wearing earplugs, and nobody's ears ruptured, but a car nearby had its windows completely blown out."

The sonic boom for Chicxulub would have been utterly mind-boggling, and any modern analog used to illustrate its scale ultimately fails. The largest nuclear weapon ever tested was the 50-megaton Soviet behemoth, the Tsar Bomba. The Tsar Bomba exploded in Siberia in 1961 and broke windows in Finland. Multiply that by 2 million and you begin to approach Chicxulub. In fact, if the Soviet Union and the United States both decided to unleash their entire nuclear arsenals developed over the entire course of the Cold War, in a single location, the Chicxulub impact would still be 100,000 times more powerful. But nuclear weapons do not represent the most powerful explosions in recorded human history. One of the loudest phenomena ever documented was the eruption of the volcano Krakatoa on August 27, 1883. "So violent are the explosions that the ear-drums of over half my crew have been shattered," wrote the ship captain of the *Norham Castle*, then 40 miles away from the volcano. "My last thoughts are with my dear wife. I am convinced that the Day of Judgement has come." The sound Krakatoa produced was heard 3,000 miles away (about the distance between Miami and Alaska) as a "distant roar of heavy guns," and it circled the globe four times.

Take Krakatoa and then make it happen, not two or three or ten more times, but 500,000 more times, all at once, and you're beginning to approach Chicxulub. As is the case with continental flood basalts—and contrary to the spirit of uniformitarianism—there is little in our world today that can help inform us about the qualitative horror of these ancient cataclysms.

One thing that can help, though, is Melosh's impact effects calculator, developed with colleagues at Imperial College London. I pulled up the calculator, which is available online, and plugged in the specs for the Chicxulub impact. For "distance to impact," I typed in "1,800 miles" to see what the K-T experience would have been like in Boston. The sound alone—at 92 decibels all the way in Massachusetts—would have been painfully loud. And with a 179-mile-per-hour air blast, it would have been sufficient to knock down wood-framed buildings and 90 percent of the trees in the area. All this from an impact that happened in Mexico.

Although the initial effects of the impact might have been fire, the literal fallout from the blast might have been a chilling coup de grâce, lending credence to Robert Frost's intuition that "for destruction ice / is also great / and would suffice." When the asteroid struck the sulfate-rich carbonate bank of the Yucatán, it injected enough sunshine-blocking aerosols into the stratosphere, it's thought, to dim the surface of the planet for months. If it was dark enough, not only would the reduction in sunlight have subjected the jungly world to punishing cold, but photosynthesis—the underwriter for almost all life on earth—would have been significantly disrupted. Perhaps such dimming explains the marine extinctions: the plankton nearly vanishes altogether at the

K-T. When these foundational legs are kicked out from under the bottom of the food chain, it isn't long before the mosasaurs at the top fall to their doom. For the trees and plants on land, this K-T impact winter might explain the modern dominance of deciduous trees over evergreens: deciduous trees are better able to withstand months of cold, dark privation and have enjoyed the advantage of this global pruning ever since. For our shrewlike ancestors that lived through this desolation, enduring blood-red skies, cauterizing winds, and the cruel, endless winter that followed—and watching as the last few giants lurched to their deaths—it would have been only natural to think that this truly was the end of the world. The tragic, unexpected coda for a spectacular planet.

The one consolation about the Chicxulub impact is that, thanks to an intensive search in the past few decades to catalog asteroids in our solar system with Earth-crossing orbits, we know with a high level of confidence that there's nothing in the void threatening us with imminent destruction. At least not for a thousand years. Eros, an asteroid even larger than the K-T impactor, has crossed Earth's orbit in the past (luckily while Earth was elsewhere on its path around the sun), and as its orbit evolves under the random tug of Jupiter and Saturn it will eventually become Earth-crossing again.

"But we're talking hundreds of thousands of years," said Melosh.

Then there's another celestial body, 8216 Melosh, an asteroid named after a certain impact expert.

"It's a main belt asteroid," he assured me. "It's not threatening the earth."

In the Yucatán, faltering attempts have been made to capitalize on the staggering Chicxulub crater, but it's tough to make a tourist site where there's nothing to see. Still, knowing this was ground zero for the most famous of all mass extinctions, I cobbled together an itinerary for a crater tour, starting at the rim. Though the crater itself might not be visible at the surface, there are other indirect ways to encounter the catastrophic structure. Not only did the asteroid impact profoundly alter the history of life on earth, but it's also shaped more than 1,000 years of human history in the Yucatán.

If you compare a map demarcating the outline of the 66-million-year-old impact crater to a map of Mayan ruins in the Yucatán, an unusual pattern jumps out. Sites like the last Mayan capital, Mayapán, are built exactly on the rim of this ghostly ring. And even stranger than the overlap of these two epochal sites—marking the final moments of Mayan civilization and the final moments of the age of dinosaurs—is that it's not a coincidence.

Like any civilization, the Mayans depended on reliable access to freshwater. In the Yucatán, freshwater is provided by picturesque sinkholes in the limestone known as *cenotes,* which appear as surprising, precipitous oases in the jungle. They form when whole sections of the limestone collapse, providing access to the underground rivers of freshwater that percolate through the chalky sea-rock of the Yucatán. The cenotes made Mayan civilization possible. When plotted on a map, the strange distribution of cenotes in the Yucatán reflects a much more profound disturbance in the rocks far below that is responsible for the limestone collapses in the region. Researchers surveying the archaeological sites that inevitably sprang up among these freshwater sinkholes made a remarkable discovery: Mayan society in the Yucatán traced an improbable 100-mile arc. UNESCO calls it the Ring of Cenotes. Walter Alvarez called it the Crater of Doom.

Gener, my local guide to the Yucatán, was puzzled by my desire to go to Mayapán, the last capital of the Maya. But the ancient city, which drew water from the holes created by the underlying impact structure—and that consequently lies on the rim of the crater—owes its very existence to the End-Cretaceous mass extinction.

"We will be the only ones there," he said.

Gener was Mayan, raised in a household that spoke Mayan, in a hometown, Yaxcopoil, that was littered with Mayan ruins. Even though I knew that the Maya didn't disappear when they abandoned their cities in the jungle 1,000 years ago, his biography (common to the millions of Maya still living in Mesoamerica) piqued my foreign ears—a little like hearing that someone is from Atlantis (or that the sparrow outside your window is actually a dinosaur). He was used to ferrying uninformed gringos such as me to well-trod sites like Chichén Itzá and Uxmal, megalithic ruins that showcase the former empire in all its classic glory before its equally epic collapse.

Mayan civilization once stretched from Honduras to Mexico, anchored by staggering cities that were hastily abandoned at the end of the ninth century by citizens and nobility alike, left to a romantic afterlife of vine-clad decay. The mysterious collapse of the Maya has been variously attributed to climate change and environmental degradation on the part of the Mayans themselves. Though Gener was fiercely protective of the need to preserve Mayan language and culture, he didn't spare his ancestors.

"They deforested the land, cut all the trees, and there was a drought . . . you would think we could learn from that, but we don't," he said, laughing. "The people started to not respect their own priests, their own governors, because they couldn't bring more water. They thought that the gods were angry with them."

After this collapse, Mayapán, a city of 16,000, consolidated the broken empire. The Postclassic period has been disparaged as a decadent shade of Mayan civilization at its peak, but the late capital had all the trappings of a complex society in full swing: from the ceremonial temple pyramid at its center (where pious Mayans vivisected their countrymen) to trade networks that spanned hundreds of miles. When Mayapán was suddenly abandoned in a panic in the fifteenth century and Mayan civilization went extinct, it was—like any extinction—a complex phenomenon. There was drought (the worst in 3,200 years), a cold snap (perhaps induced by volcanism on the other side of the world), and famines that prompted the murder of the entire ruling royal family and a horrified evacuation of the city. Mayan civilization might not have been doomed to oblivion at this point, but the following decades offered no reprieve. There was a massive hurricane, a plague called the "blood vomit," scattered warfare that may have killed 150,000 Mayans, and finally an event with no precedent at all. With Mayan civilization gasping from decades of environmental and societal chaos, the end came for them sailing silently over the horizon—a final insult from beyond, as unforeseeable as an asteroid impact. In 1517 the Mayans met the Spanish. No more pyramids would be built.

This is how societies, ecosystems, and, it turns out, whole worlds collapse. As we'll see, a subtler picture is emerging about the death of the dinosaurs that hints at there being more to the catastrophe than a single, terminal throttling. Like the end of Mayan civilization, it might have been a series of increasingly improbable blows that brought the Cretaceous to its dreadful end. If you flip a coin enough times, it will eventually come up tails 100 times in a row, and the earth is very old. This was as true for dinosaurs as it was for the Mayans as it is for our modern global society.

My crater tour had still one more stop: ground zero. Gener and I drove from the crater's rim in Mayapán to Chicxulub Puerto, the town in the crater's very center. This jaunt took more than an hour of driving on the highway and covered only the crater's radius, providing some scale for the catastrophe. Chicxulub is hallowed ground in the history of life on earth—a place name that has become synonymous with the epic death of the dinosaurs. The world is filled with tourist attractions commemorating historical events that have taken place in the eyeblink of our own species' recent history, but here in Chicxulub was the site of a historical event an order of magnitude more important: a celestial holocaust that rerouted the trajectory of life on earth and made our very existence possible.

After passing through a military police checkpoint, we found Chicxulub itself to be a colorful beach town, lined with low-slung open-air cafés serving snapper straight from the ocean and ice-cold Sol beer. On the beach, I looked out over the toothpaste-blue water patrolled by seabirds and tried to imagine the last moments of the Mesozoic. Some of it I could vaguely comprehend, like the brief, ghostly limn of the asteroid appearing suddenly in the daytime sky like an irregular moon. Still out in space, it was on the same airless vector as it had been for eons, a continent of space garbage about to terminate on the most interesting place in the solar system. These were the last tranquil moments before the end of the Mesozoic, as pterosaurs flew blithely over the surf, scanning the shallow sea for fish, unaware of the peculiar pale figure growing in the sky. It is possible, but less likely, that the offending rock was an icy comet and not an asteroid. In that case it would have heralded the end-times somewhat more dramatically, blazing in the sky for weeks like a chariot of Death. As night fell in the final weeks of the Cretaceous, *hadrosaurs* settling down for a fitful sleep would nervously eye this strange new star, which cast midnight shadows onto the forest floor. Hundreds of millions of years of dinosaur lives stretched behind them, and yet only a few precious hours remained. The unfamiliar beacon would have been beautiful but strangely unsettling, stretching across half the sky. *Tyrannosaurus rex* (it's easy to forget) was an actual living, breathing animal. It would have similarly witnessed this spectacle, which in the days before landfall would have been a striking feature of the daytime sky as well. These scenarios I can imagine, if only barely. What followed this silent approach, though, beggars all powers of imagination.

In the colonial Massachusetts town in which I grew up, every other clapboard-sided house has a plaque on it, commemorating the life of some eighteenth-century minor cabinetmaker or other. Chicxulub Puerto cannot be said to suffer from the same sort of historical egotism. Here, hidden behind carnival tents and fair games assembled in the town plaza for Easter vacation, was a strange stucco monument. The concrete slab was faced with goofy dinosaur skeletons carved in relief. At its base was a broken, bone-shaped blob of cement with an epitaph of odd, misspelled Spanish scrawled, apparently with a penknife, and memorializing the CATACLISMO MUNDIAL.

AN ENORMOUS ASTEROID 10 KM IN DIAMETER HIT IN THIS LOCATION, it read.

It was ground zero for the most important event of the last 100 million years and this was its only memorial.

"It looks like it was made by kids," said Gener.

"Chixculub is not all that it's cracked up to be," says Princeton geologist Gerta Keller, the world's biggest skeptic of the Alvarez Asteroid Impact Hypothesis.

The barb, delivered in a vaguely Teutonic accent, was typical of Keller's sly needling of the status quo. She doesn't doubt that a big rock hit Mexico sometime near the end of the Cretaceous, but she thinks the idea that it caused the mass extinction is preposterous.

Hanging above her desk in Princeton, New Jersey, is a very different artist's rendering of the end of the age of dinosaurs than the Armageddon-by-asteroid, an extinction scenario now so widely accepted that it merits only a brief, knowing mention in any story about dinosaurs. Keller thinks it's bunk.

Lethal heat pulse?

"Insane. And there's no evidence for it."

Nuclear winter?

"Nah."

The tsunami deposits in Texas?

Nonsense.

The image of Caribbean tyrannosaurs squinting at the celestial apocalypse on the horizon, moments before being blown apart, has become a fixture of the natural history museum, and almost part of pop culture. And undoubtedly any dinosaurs that were anywhere near the impact would have surely witnessed what looked like a tropical Judgment Day, before a swift, blinding, and painless death by evaporation. But Keller scoffs at the idea that a local event like an asteroid impact could bring an end to the Mesozoic.

"It seems to me entirely fantasy to think that if impacts leaving craters of 100 to 120 kilometers don't do anything, that one that's 150 to 170 kilometers is wiping out the whole earth," she said, implicitly comparing duds like the Triassic Manicouagan crater in Quebec to supposed world-destroyers like Chicxulub.

In the picture above Keller's desk there are two tyrannosaurs, somewhere in India. In this artist's depiction, they lay frothing at the mouth, writhing in agony. In the distance is a landscape riven by towering volcanoes and gashes in the earth, vomiting incandescent floods of lava. In the foreground, the landscape is arid and withered. Trees and shrubs are rotting, and the air is a sulfurous haze. As with the End-Permian and End-Triassic mass extinctions,

Keller said, volcanoes brought an end to the age of dinosaurs. It was an end-time of global warming and ocean acidification, just like the Permian and Triassic apocalypses that came before.

That Keller can say this is not merely testament to the enduring spirit of contrarianism. Amazingly, at almost the exact same moment that Earth was struck by the largest asteroid known to have hit in a billion years, western India was being smothered—in some places more than 2 miles deep—in lava. So profound was this Indian volcanism that it would have been enough to cover the entire lower forty-eight United States in 600 feet of lava.

This is the most confounding aspect of the K-T. And it's one that, for thirty-five years, has been more than a little irritating to those who have advocated for a simple asteroid story to explain away the mass extinction. As the frantic search for asteroids at other extinction boundaries has turned up only flood basalts, the curious existence of one of the largest flood basalts *ever* at the very end of the Cretaceous remains the fly in the ointment for what otherwise seems to be a nice, neat story about an asteroid impact. Although she's not the first to do so, Keller's insistence that these Deccan Traps, not the crater in the Yucatán, brought an end to the Cretaceous has won her few friends in the K-T community.

Keller did not chart the typical path to the rarefied position of tenured geologist at Princeton. The black sheep and youngest of twelve in a rural Swiss farming family, she repeatedly had her ambitions thwarted, first as a teenager by a shrink who told her to abandon her hopes of becoming a doctor (she came from too meager a background for such daydreams), then in a drudgery-laden apprenticeship as a dressmaker. Determined to see the world, the restless Keller did the only sensible thing: she picked up and hitchhiked across North Africa and the Middle East.

"I'm a strange bird," she admitted.

While preparing for a journey on the trans-Siberian railroad, she became deathly ill while caring for a sick fellow traveler.

"I got so sick I took a train to the Vienna hospital, where they thought it was a miracle still to be alive," she said. "I was isolated for six weeks on intravenous before checking myself out of the hospital. I have a habit of doing that."

Six months later, she ended up in Australia.

"Then I got shot in a bank robbery," she said unemotionally. "I was pronounced dead."

A bullet passed between her heart and spine, pierced her lungs, and shattered her ribs on the other side.

"The entire blue sky was essentially a film of my life going by, and by the time I got to the end I said, 'I don't want to die,' and that was it. I had this weird experience where I came to, and I was floating over Sydney. It was very peaceful. I was watching a park and the people in the swimming pool in the park, and the cars going around it. Then there was this ambulance and siren breaking this peace, making a big racket. I heard this woman scream and calling for her mama and I got really annoyed. I thought, 'Jeez, I would never do that.' Then all of a sudden I was sucked down and I was that woman screaming, and there were two orderlies holding me down."

I asked whether she thought she actually did fly over Sydney or whether her visions were the hallucinations of a brain under extreme duress.

"It was real!" she protested. "I saw areas I had never seen that I later went to. It was very weird.

"The funny thing is, I always thought I would die at twenty-three," she continues. "I didn't want to get old, and when I was young I put my limit at twenty-three. I was shot at twenty-two, and then I decided maybe it's not so bad to be twenty-three."

Like Peter Ward's near-drowning, Keller avoided extinction by the narrowest margin.

Her life continued its improbable trajectory. She landed as an adult high schooler in the ghettos of San Francisco before bouncing to San Francisco State, where her academic life finally assumed a more traditional path.

This hardscrabble background might explain Keller's toughness (some might say stubbornness) in the academic arena, where she almost gleefully invites the scorn of her colleagues by pushing against the prevailing K-T narrative. She told me that she still has a shoebox of letters mailed to her by one particularly irate academic rival who sent her ten-page, single-spaced screeds every week for five years. Her effectiveness in exasperating her colleagues is evident in the Q&A sessions after her talks at academic conferences, which resemble less an open exchange of ideas than a turf war.

Keller points to what she thinks are wild carbon dioxide–driven swings in climate at the K-T, deciphered from the isotopes of fossil plankton pulled up in drill cores from the deep ocean off Namibia, as well as in Tunisia and Texas. Rapid warming spikes of 4 to 5 degrees Celsius in the ocean, and as much as 8 degrees on land, that took place in under 10,000 years decimated life, she said. Meanwhile, just as in the extinctions of yesteryear, the oceans acidified, as evidenced by the dwarfed disaster species of plankton that thrived for thousands of years after the extinction.

"This is what a normal assemblage looks like," Keller said, producing a picture of the intricate spiral shells of single-celled plankton that thrived before the extinction. "Then when we introduce a stress, climate change, the specializers drop off. If you keep increasing the stress, and you get to super-stressed conditions, this is what it looks like. This disaster opportunist takes over while everything else is going extinct."

Over the extinction boundary the organisms' shells became smaller, simpler, and uglier—reduced to little blobs of calcium carbonate. Winnowed by ocean acidification and extinction, the surface area of the shells shrank as a bulwark against corrosion.

"These are the guys you can't kill."

Just as with an impact winter, ocean acidification provides a compelling mechanism to take out the bottom of the food chain and send the entire marine ecosystem toppling over, mosasaurs and all.

Keller thinks the momentum is on her side. Layers once interpreted as soot from global forest fires ignited by the asteroid ejecta have been challenged, while the oxygen requirements of such a global fire might make this proposed kill mechanism dubious to begin with. Claims that the atmosphere was temporarily heated to the temperature of a pizza oven has long been viewed with skepticism by biologists, Keller said, as birds, mammals, amphibians, and reptiles that breathe air survived, which they tend not to do in pizza ovens. As for the acid rain–causing, sunlight-blocking sulfur dioxide theorized to be a major component of the Chicxulub impact, Keller's colleague Anne-Lise Chenet at the Institut de Physique du Globe de Paris estimates that just a single large pulse of Deccan volcanism would have injected the same amount of this smog into the atmosphere as the total output of the asteroid impact. The kilometer-high tsunamis racing around the world originally envisioned by impact proponents have been revised down to only a few tens of meters—still terrifying, but not sufficiently so to be an agent of the apocalypse.

It's Keller's interpretations of the same tsunami deposits I saw in Texas, and elsewhere in Mexico, that form the tenets of one of her chief heresies. To Keller, Smit's tsunami layer on the Brazos River outside Waco represents not a catastrophic sand dump after a killer wave, but the peaceful, humdrum deposition of a sandy sea bottom, laid down over many thousands of years. There's evidence of the Chicxulub impact at the base of the rocks, Keller said, but in the sandstone layer above she pointed to critter burrows (either disputed by or invisible to her rivals), which indicate that this was long a peaceful environment settling out over at least 100,000 years—not one created in hours or days by destruction. Based on her interpretation of fossil plankton in these rocks, Keller puts the mass extinction at more than a meter above this controversial sandstone, meaning it occurred long after the impact and thus could not have been caused by the asteroid.

It was an interesting explanation, I said, but admitted to Keller that the iconic Alvarez story about Gubbio—the plankton extinction, the sudden break in the limestone, the clay layer, the iridium discovery—well, it all just seemed so damn *persuasive*. She laughed heartily.

"It's very persuasive, yes, if you know nothing about stratigraphy."

Keller claims that the famous Alvarez section in Italy compresses millions of years and is missing huge sections of time. In addition, she disputes a 2013 finding by Berkeley geochronologist Paul Renne, widely regarded as the most authoritative and precise dating of the impact and the extinction to date. Renne, studying the world-famous K-T deposits in Hell Creek, Montana, was able to place the asteroid strike and the death of most animal life on earth within 30,000 years of each other—about as close as the geologic record allows. But Keller stands by her finding that the impact predated the extinction by at least 100,000 years, claiming that the error bars in Renne's work leave wiggle room.

"That's just complete nonsense," Renne told me in a phone conversation, echoing the reaction Keller typically receives when presenting her results. "You know, Gerta's her own worst enemy because, on the one hand, I think she's right about there being a connection to the Deccan Traps. She's really pursued it, and I'm sure she's done some really important work on that front. But this insistence that Chicxulub predates the K-T boundary by—you know, initially it was 300,000 years, and she's been shrinking it lately—but it just completely undermines her credibility. She's stretched geological reality in ways to accommodate her views that are just absurd."

One area of Keller's work that is not disputed, however, is her team's recent results in dating the Deccan Traps themselves. Each field season Keller's group, led by her Princeton colleague Blair Schoene, has shipped off to India to try to tease the ages out of these staggering lava piles from the end of the Cretaceous. With each new date the Princeton team gathers, the closer and closer the most destructive period of volcanism inches toward the K-T boundary.

More than 100 million years ago, India divorced the supercontinent Gondwana, a long-standing marriage that had lasted since the beginnings of animal life. India was evidently unsentimental about this union, hightailing it across the proto–Indian Ocean in the late Cretaceous at a blistering (geologically speaking) pace of almost a half-foot a year. But before its terminal rendezvous with Asia, India had the misfortune of wandering over the Réunion Hotspot. Sometime near the K-T, the island continent briefly turned inside out, erupting on an unimaginable scale.

Amazingly, the Deccan Traps are still erupting today. On the eastern flank of the island nation of Réunion, 500 miles east of Madagascar, the Réunion Hotspot that once gushed from India still gushes lava. The Piton de la Fournaise volcano, and indeed the entire country of Réunion, is just the latest expression of this anomalously hot section of earth's mantle—one that has a hoary legacy dating back to the end of the Cretaceous. Like the Hawaiian Islands, which get younger from west to east as the Pacific plate slides over an angry plot of mantle far below, the Réunion Hotspot has similarly traced a path across the Indian Ocean as the plate shifts overhead. In the 66 million years since the eruption of the Deccan Traps, the hot spot has pierced the crust to create the Maldives, the Seychelles, and Mauritius before arriving at its present location underneath Réunion. But this hot spot first began to seethe under India.

Hundreds of thousands of years before the K-T mass extinction, with dinosaurs still happily trudging the globe, this section of mantle first awoke and the lava came out in spurts, confined mostly to a small region in and around Mumbai. Though these lava floods would appear catastrophic to us today, Keller said that these initial eruptions account for only 4 percent of the lava ultimately erupted by the Deccan Traps. Nevertheless, even these earlier, relatively minor eruptions of the Deccan Traps seem to have wreaked havoc on the climate. First there was a sharp warming spike, evidence of ocean acidification, and a swing in the carbon cycle 150,000 years before the extinction, thought to be related to the injection of volcanic CO_2. These blushes of warmth might have been followed by an icy plummet. Plant fossils from North Dakota show temperatures dropping as much as 8 degrees Celsius in the geological moments before the mass extinction. This freeze might have been caused by the weathering of the fresh Deccan basalt from the first phase of eruptions and an attendant drawdown in CO_2. The chill might have even brought glaciers to what was once a greenhouse world, as sea levels appear to plummet right before the K-T boundary. Working in Montana's legendary Hell Creek badlands, which span the End-Cretaceous mass extinction, University of Washington paleontologist Greg Wilson has documented the local extinction of about 75 percent of mammals in what looks like the geological moments *before* the asteroid strike, with the once-dominant group that includes marsupials getting absolutely hammered: ten out of eleven species went extinct. At the very least (if these signals are real), such profound shifts in climate and fauna seems like an odd prologue to a completely unrelated, asteroid impact–driven Armageddon.

Then again, some are skeptical that the pre-K-T temperature swings were all that extreme, or that the biosphere was anything like on death's door before the asteroid struck. In a comment in the journal *Science,* Penn State paleobotanist and impact proponent Peter Wilf colorfully derided this "bombing-the-nursing-home scenario."

"The killer was caught long ago," he wrote in an email to me, referring to the asteroid.

If it was the Deccan Traps and not the Chicxulub impact that brought an end to the Mesozoic, it probably would have occurred during a cataclysmic pulse of eruptions in the volcanoes' middle age. Sometime after the initial spurts, at around the same time as the asteroid impact, the volcanic system in India suddenly went from a sputtering fissure to a broken fire hydrant of molten earth. This "main phase" of the eruptions covered an area the size of France more than 2 miles deep in lava in some places.

Today in western India, 11,500-foot-tall bar-coded mountains, like the jagged, banded basalt peaks of Mahabaleshwar, have been carved from this surfeit of molten rock. Ancient lava from the Deccan Traps is even found spilling over off the other side of the subcontinent, into the Bay of Bengal, transported there by "the most extensive and voluminous lava flows known on Earth." These molten rivers carried about 2,400 cubic miles of lava over a distance of perhaps almost 1,000 miles, or roughly the distance between Chicago and Boston.

And just as the Mayans unwittingly alighted on the dinosaur-killing crater in Mexico—relying on it for freshwater—on the other side of the world Buddhist monks also found the geology of the End-Cretaceous mass extinction agreeable. Beginning twenty-two centuries ago in the jungles of the Western Ghats, Buddhists carved some two dozen monasteries and five temples into the cliffs of this Deccan basalt. These are the stunning Ajanta Caves, which, like the Ring of Cenotes, are a UNESCO World Heritage Site. Thousands of years ago, monks sat in silent meditation deep inside the rock that might have destroyed the world, a fitting place to contemplate that foundational Buddhist concept of *anicca*—the impermanence of all things.

At the annual meeting of the Geological Society of America, UC Berkeley geologist Mark Richards took the podium to address a packed room, many of them battle-scarred veterans of the past thirty-five years of debate on the K-T. The crowd had just listened to a series of talks by Keller and her colleagues dismissing the Alvarezes' asteroid, placing the blame squarely on the shoulders of Deccan volcanism. The lines of the battle were once again clearly drawn, and Richards's talk was billed in the program as a collaboration with K-T superstars Walter Alvarez, Jan Smit, and Paul Renne, who had done more than anyone else to date the asteroid impact almost precisely to the time of the extinction. If there was ever a dream team to refute the annoying paleo-fantasies of Gerta Keller, this was it.

But Richards immediately began his talk on an unusual note.

"I want to start off this talk by talking about an 800-pound gorilla that has kind of turned into an 8,000-pound gorilla in my view."

In past meetings, scientists who vouched for the lethality of the asteroid mentioned the existence of the Deccan Traps only in dismissive asides, and vice versa. But Richards tackled the issue head on, acknowledging the bizarre (and for some, uncomfortable) coincidence between the biggest asteroid and one of the biggest flood basalts in the entire eon of animal life.

"The chances of this occurring at random seem to either imply some sort of causal effect or some kind of divine intervention," he said. "And since I'm not an expert on the latter—I'm a geophysicist—I'm going to concentrate on what might have been considered kind of an outrageous hypothesis a couple of years ago, but which I no longer think is such an outrageous idea."

Like Keller's group, Richards, Renne, and others had been spending time in India, poking around the lava piles. After perilous journeys up and around the Western Ghats, sharing cliff-hugging roads with daredevil motorists, the group sampled rocks from the staggering peaks of Mahabaleshwar. Richards and company were especially curious about an apparent break in the rocks partway up the lava pile. After the first few spurts of lava at the base of the Traps, something fundamentally changed in the rocks. This was the beginning of the Wai subgroup, a monstrous stack of lavas within the pile that accounts for at least 70 percent of the entire Deccan Traps.

"If you took out the Wai subgroup, [the Deccan Traps] wouldn't even be considered a world-class flood basalt event," said Richards.

Not only is the volume of lava in the Wai subgroup unusual, but the chemistry of the rock itself changes compared to the rock below. In the smaller eruptions earlier in the history of the Deccan Traps, elements from earth's crust are generously mixed in with the lava, indicating a leisurely ascent to the surface. But at the start of the Wai subgroup, the lava bears only the signatures of rock from deep inside the earth, signifying a wild, precipitous escape from the deep and almost no interaction with the crust.

Ediacaran fossils at Mistaken Point, Newfoundland, Canada. The frond-like creature imprinted in this 565-million-year-old rock would have stood upright at the bottom of the ocean at the dawn of complex life, absorbing nutrients across its membrane. Strange, immobile life forms like this dominated in the ocean until the Cambrian explosion, when animals rapidly diversified and wiped all of them out.

Life in the shallow seas that covered much of North America during the Late Ordovician, featuring nautiloid cephalopods, trilobites, crinoids, byrozoans, brachiopods, and early fish. *© 2003 Douglas Henderson,* Ordovician Marine, *commissioned by Museum of the Earth, Ithaca, New York*

An outcrop of rock from the bottom of the Ordovician ocean in Southwest Wisconsin. Weeds are growing in layers of volcanic ash.

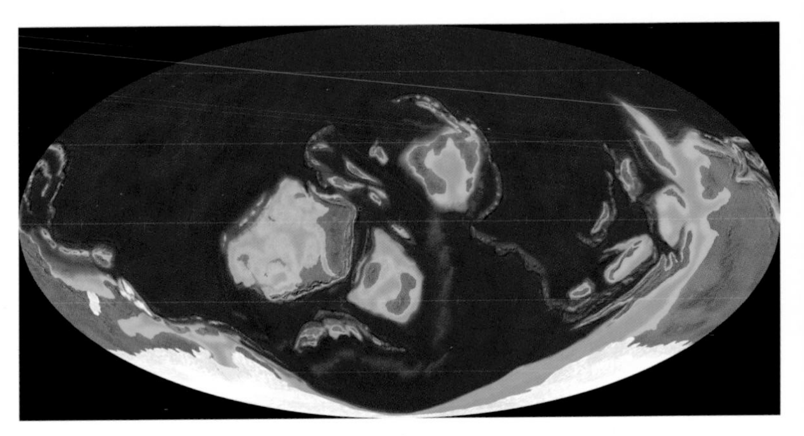

The Late Ordovician world. North America straddles the equator and is rotated almost 90 degrees with the early Appalachians forming a range on its southern coast. Most of the continent is covered by a shallow sea. © *2016 Colorado Plateau Geosystems Inc.*

The strange layers of the Late Devonian mass extinction exposed on the shores of Lake Erie. These black shales are filled with hydrocarbons from ancient organic matter sinking to the seafloor in the Devonian. Black shales are found around the world associated with the Devonian extinctions and are indicative of widespread ocean anoxia.

The head armor of Late Devonian apex predator *Dunkleosteus*. Placoderms like *Dunkleosteus* were extirpated by the mass extinction at the end of the Devonian. Photo taken at the Rocky River Nature Center outside Cleveland, Ohio.

The Permian gorgonopsid therapsid, *Lycaenops*. Therapsids like this were wiped out, along with most other animal life, at the End-Permian mass extinction.

El Capitan, Guadelupe Mountains, Texas. This limestone promontory is, in fact, a 260-million-year-old coral reef from the Permian. The sea life that made up this ancient reef—species of sponge, corals, brachiopods, crinoids, fusilinids, ammonoids, etc.—was driven almost entirely extinct by the end of the Permian.

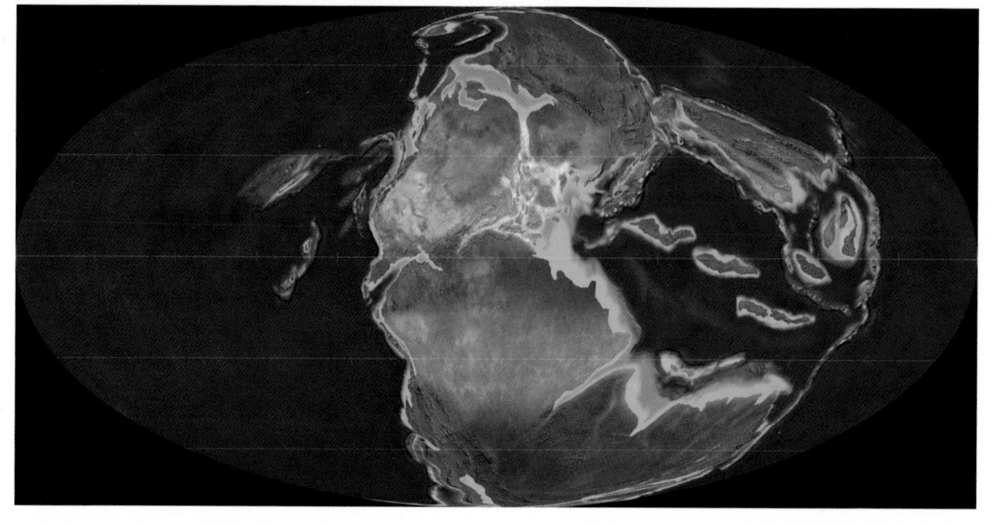

Pangaea in the Triassic. The Siberian Traps erupted in northern Pangaea and are visible in this representation as a dull gray-brown swath of Siberia at the top of the globe. The interior of the supercontinent, from present-day Ireland to the western United States, was, for much of Pangea's history, extremely unpleasant. © *2016 Colorado Plateau Geosystems Inc.*

Solite Quarry on the border of North Carolina and Virginia. One of the most important Triassic fossil sites in the world, the layers are from an ancient lake bottom, when a rift valley stretched from North Carolina to New York, and the east coast of the United States was contiguous with western Africa.

The Carolina Butcher, a Triassic bipedal crocodilian discovered not far from Chapel Hill, North Carolina. Distant crocodile relatives like this, and a variety of other bizarre forms, ruled the world in the middle to late Triassic. © *2015 Jorge Gonzales*

The New Jersey Palisades along the Hudson River, part of the Central Atlantic Magmatic Province (CAMP) continental flood basalt. These cliffs are made of magma that fed the eruptions associated with the End-Triassic mass extinction. © *Jim Wark/Visuals Unlimited, Inc.*

Life in the ancient rift valley of today's Connecticut River Valley 210 million years ago, before the End-Triassic mass extinction. The dominant predators of the time are not the small dinosaurs (as seen to the right) but an array of crocodile relatives. © *William Sillin/Dinosaur State Park*

Ammonites. Animals like these swam in the seas for hundreds of millions of years, from the Devonian until their extinction at the end of the Cretaceous, with some housed in shells upwards of eight feet in diameter. © *2015 Simon Stålenhag/ Swedish National History Museum*

Tyrannosaurus rex tearing apart a carcass while a flock of nervous herbivores skitter by in the foreground. Tyrannosaurs were mostly small and unimportant for the better part of their 100-million-year history. But in the final 20 million years of the Cretaceous before the mass extinction, some species like *T. rex* reached outrageous size and fearsomeness. © *2015 Simon Stålenhag/Swedish National History Museum*

Tyrannosaurus rex, with the Chicxulub asteroid hovering in the sky, moments before the catastrophic impact that would have released, all at once, far more energy than all the nuclear weapons ever detonated during the Cold War. © *2000 Douglas Henderson "T-Rex & Asteroid" from Asteroid Impact, pub by Dial, 2000*

Digging in Paleocene rocks at Angel Park Scenic Area, New Mexico. By piecing together the record of plant and animal fossils, as well as geochemical signals from the rocks pointing to ancient climate changes, paleontologists can reconstruct how the planet recovered in the aftermath of the End-Cretaceous mass extinction.

Deccan Traps in Mahabaleshwar, India. These mountains are carved entirely out of ancient basalt lava. This vast volcanic province erupted around the same time as the End-Cretaceous mass extinction with enough lava to cover the continental United States in lava 600 feet deep. © *Gerta Keller*

Dinosaur memorial at the epicenter of the asteroid impact in Chicxulub Puerto, Yucatán, Mexico.

"This is like a fire hose right out of the mantle," Richards said.

While Richards was unpacking his data from India, his good friend, Berkeley colleague, and paleontological god, Walter Alvarez, was studying the geography of the Traps on Google Earth. He noticed in satellite photography that the fault lines that run through the weaker, older lava flows don't extend through the enormous Wai subgroup on top. To Alvarez, this indicated a significant break in time between the earlier eruptions and the fire hose. Alvarez excitedly phoned Richards with his armchair revelation.

"When you're working on problems that involve the K-T boundary and Walter Alvarez calls you up on a Sunday afternoon and says you need to come over right now, you need to come over right now," said Richards. "What this suggested was that the Deccan Traps were yawning and maybe saying, 'We're done,' . . . and then something happened.

"Here's the kicker," Richards said. "A group of paleontologists, including Gerta Keller, showed that if you go and look at drill cores over in the Bay of Bengal where these huge lava flows cap Cretaceous sediments, those lava flows come in exactly at the Cretaceous-Tertiary boundary."

Richards's gestalt moment came where else but on a family vacation to the Mayan ruins of the Yucatán. Before leaving, Alvarez showed Richards the curious map illustrating the crater in the Yucatán with the cenotes that powered Mayan civilization, apparently hugging the rim of the crater.

After he visited a cenote near Chichen Itza, inspiration struck Richards.

"I came back to my hotel room, and I had about as close to what you would call an epiphany as you get in this business," he said. "I literally sat up, bolt-straight in my bed, at three in morning and got out my computer while my family was asleep and started searching the literature."

Richards suddenly recalled the work of his UC Berkeley colleague Michael Manga, who had worked on the hypothesis that earthquakes could trigger distant volcanoes. It wasn't exactly a new idea, but it was one that was beginning to gain the force of statistical validation.

In 1960 the largest earthquake in recorded history struck Chile. Thirty-eight hours later, the Cordón Caulle volcano blew its top 150 miles away. More than a century earlier, Charles Darwin experienced a similar earthquake in Valdivia, Chile, and within a day came the eruptions of Minchinmávida and Cerro Yanteles. Darwin sensibly imputed a causal connection between the phenomena, but struggled to come up with a convincing mechanism. For many years, this suggestive evidence for the intuitive link between earthquakes and volcanoes was only anecdotal. But recently the application of statistics has revealed it to be a real phenomenon. It seems there is a relationship between the magnitude of an earthquake and the distance at which it can trigger volcanoes. It was this relationship that Richards's colleague Manga was sketching out. Scaling up to a preposterous magnitude 11 earthquake, of the sort caused by Chicxulub, Manga calculated that the triggering distance is effectively global. That is, a magnitude 11 earthquake should be capable of setting off volcanoes all around the world. An earthquake of the size induced by the Chicxulub impact—Richards realized in his witching-hour revelation—could have turned the humdrum volcanoes of India into an agent of the apocalypse.

In 1997, while making the case for Chicxulub, Alvarez himself puzzled at the almost miraculously coincidental timing of the asteroid and the volcanism, writing: "A good detective shouldn't ignore even a single coincidence like the K-T Deccan match in timing, and when it is bolstered by a second coincidence like the match between the Siberian Traps and the Permian-Triassic boundary, it just has to be significant. But at the moment, I don't know of anyone with a reasonable explanation for a link between impacts, volcanism, and mass extinctions."

Now they had their link.

The idea that the asteroid impact might have caused the volcanism in India wasn't entirely new. Briefly entertained in the decades before, it had been abandoned when it was determined that the initial pulse of the Deccan Traps predated the impact, possibly by millions of years. If the volcanism started before the asteroid hit, researchers reasoned, there was no hope for any causal connection. The Deccan Traps also weren't antipodal to the impact, where the seismic energy would have been focused on the far side of the globe, and in any case, the impact wasn't strong enough to cause such a catastrophe on its own even if it had been.

But if Richards's idea is right—if Chicxulub could have caused a volcanic system that was already waiting around for an extra kick to explode—he suspects that not only would the Deccan Traps have been kicked into overdrive by the impact, but so too would arc volcanoes all around the world, along with the entire midocean ridge system. After the impact, the jigsaw seams of the planet would have lit up in a constellation of volcanic cannonades.

Despite the possibility of a thrilling reconciliation between warring theories and a revival of the role of volcanism in the End-Cretaceous mass extinction, Richards has wisely avoided assigning primary responsibility for the death of the dinosaurs. The two phenomenal events still hover uneasily close to each other, somewhere very near the extinction. He's evasive when pressed on the ultimate role of the Deccan Traps in the extinction as new, finer-scale measurements continue to pour in from the field about the nature and timing of the eruptions.

"The best decision I ever made was to take absolutely no opinion about what caused the K-T boundary extinction," he told me. "There's just no reason to make a statement like that. I mean, this has been a very acrimonious debate. I've been trying to steer clear of that. We're going to find out in the next couple years what happened. We might as well just all be friends and find out. Together. This story's only going to get more interesting."

His colleague Paul Renne is not as shy.

"I think maybe we are moving away from fireballs and damnation and that sort of thing," he said.

"It's been a great mystery for many years as to why in the world—where no other mass extinction that we know of is definitively related to an impact and, by contrast, the bigger ones are all associated with flood basalts like the Deccan Traps—why this bizarre coincidence?

"It may be that Chicxulub was the gun and the Deccan Traps were the bullet," he said.

So what would it take to wipe out the dinosaurs, the most dominant animal group on land in the history of the planet and one that ruled the earth for 136 million years? Well, this might do: a climate that was deteriorating at the end of the Cretaceous, with greenhouse heat waves punctuated by brief and bitter winters . . . interrupted by an asteroid the size of San Francisco plowing through the atmosphere in a second, creating Mordor in Mexico, incinerating everything around, sending tsunamis hundreds of miles inland over distant shores, collapsing the eastern seaboard, bringing an age of darkness, failing plankton blooms, foundering food webs, and acid rain, and then . . . on the other side of the world, while the midocean ridges grumbled at the bottom of the ocean, the earth opened up, as it has in only a few catastrophic chapters previously in its history, drowning western India in fire, acidifying the oceans, and bringing punishing warmth to the world for thousands of years.

Of course, this remains a speculative sketch. The fact is, we still have no idea what the last days of the dinosaurs were like. The only thing we know is that they were unspeakably awful.

It seems reasonable to ask how dinosaurs could have been dealt such an improbably terrible hand. The sheer overdetermination of kill mechanisms in their death seems almost to hint at some vengeful, dinosaur-hating destroyer god. More likely, it was an unfortunate consequence of being so successful. Dinosaurs absolutely dominated the planet for a functional eternity. The longer you're around, the more likely you are to see some very, very rare, very, very bad stuff. Humans have been around for far less than a million years, but if we can hold on for a few *hundred* million years more, we too will have some good days and some bad days.

I had just spent a long day driving around rural Alabama with University of Alabama paleontologist Dana Ehret, hunting for mosasaurs, and I was eager to call it a day, go home, and nurse my sunburn. We had a good haul: finding vertebrae from the fallen monsters was easy. They wash out of the state's eroding gullies of chalk after each rainstorm. But Ehret wasn't done. He wanted to hunt for a local outcrop of the storied K-T boundary itself, the line in the rocks across which no mosasaurs swim. Ehret calls himself "the Alabama Rambler," and he was living up to his name.

"Where the hell are we?" he asked.

The rivers were running 8 feet higher than normal that day, so his go-to K-T outcrop an hour or so from campus was flooded. He had a plan B, though, a spot he'd never visited. A retired Mississippi state geologist had tipped him off to it, like a fisherman entrusting a favorite spot to a fellow sportsman.

"We are in the middle of nowhere," he said, fumbling with a paper map, our phones now out of range and useless on the Alabama backroads, two and a half hours south of Tuscaloosa. We passed cotton field after cotton field, miserable rusted tin shacks, and a shirtless man in overalls, sitting on his front porch with a gun. Ehret looked back at the map and then up again at an apparently unexpected intersection.

"We are in the middle of nowhere," he said again, pausing.

"Let's just go a little bit farther," he added. "It's too exciting."

We were getting close. The road's shoulder was littered with oyster shells from the very end of the Cretaceous—huge gnarly seashells nicknamed "Devil's Toenails." If we drove too far south, we'd zoom past the K-T and, instead of mosasaurs and tyrannosaurs in Alabama and Mississippi, we'd find giant whalebones and early primates.

We passed a pickup truck hauling hundreds of crestfallen chickens stacked in rusty metal cages, the highway wind ruffling their feathers. Some of the humiliated dinosaurs glared at our van—with those unmistakably reptilian eyes—indicting us in the indignities foisted upon their once-proud lineage.

680 ACRES FOR SALE, a ragged country sign on the side of the road read.

"Six hundred and eighty acres right on the Cretaceous-Tertiary boundary!" Ehret joked. Suddenly a great, stratified edifice surged from the highway shoulder.

"That may be it," he said, abruptly serious as he craned forward for a better view. Ehret hastily pulled the van over at high speed into the weeds and parked, resting on the steering wheel. Halfway up the roadcut, the rocks abruptly switched colors.

"That's definitely it."

At the top of the cliff was a new world. Our world.

7

THE END-PLEISTOCENE MASS EXTINCTION

50,000 Years Ago—Near Future

The world that followed the endless age of dinosaurs, the Paleocene, was a weird one. In the canyons of Angel Peak, New Mexico, Williamson's team was trying to reconstruct this shell-shocked new planet. These New Mexican badlands were chock-full of turtle shells, gator bones, and mammal teeth. Familiar enough, but most of these mammals were barren lineages with no descendants in the modern world. Elsewhere, the earth was running truly strange experiments, straining to fill the ecological chasm left by the dinosaurs' disappearance. In South America, there was *titanoboa,* a 2,500-pound snake stretching almost 50 feet. The monster snake would be matched in fearsomeness by the continent's "terror birds," which first evolved in the Paleocene but would later grow heads the size of horses, dinosaurian feet, and giant hooked beaks, with which they terrorized the countryside, carrying on the family business of their dead cousins.

"It's wild," Stephen Brusatte said. "There were some very weird types of birds that essentially filled dinosaur niches. Of course, birds *are* dinosaurs, but you know, filling the niches left by things like velociraptors and stuff, during the few million years after the extinction."

If the fauna was volatile, the climate was even less predictable.

"The Paleocene and Eocene was a time of very chaotic swings in climate," Brusatte said as we hiked under a withering New Mexican sun over dry, dusty streambeds. This desert heat was nothing, though, compared to the global sweat lodge faced by our ancestors.

"We know it was a really hot time, much hotter than today. And given where we're headed, we want to know what our planet's going to be like during hot times. Not only was it a lot hotter, but you have these big spikes in temperature and they last for maybe tens of thousands, or a few hundred thousand, years at most. That's why we're here studying it."

"I found a complete turtle plastron!" Williamson shouted from atop the badlands.

"Oh, cool," Brusatte shouted back, surveying the imposing ridge. "How'd you get up there?"

The hothouse of the early age of mammals hit its sweltering maximum 56 million years ago, when an amount of carbon roughly equivalent to today's fossil fuel reserves was released to the atmosphere and oceans over the course of less than 20,000 years. As a result, the temperature spiked 5 to 8 degrees Celsius. This is known as the Paleocene-Eocene Thermal Maximum, or PETM. The source might have been volcanoes deep in the North Atlantic, burning through huge stores of fossil fuels under the seafloor. As the carbon dioxide and methane degassed the climate would have sizzled, perhaps kicking off a feedback loop by thawing permafrost on land, which then would have released ever more carbon dioxide and methane, warming the planet even further. None of this should sound encouraging to modern ears.

Coral reefs took a grave body blow in the PETM, while mammals, like early horses, shrank in size to beat the heat and raced poleward, where the Arctic Ocean was a tepid 76 degrees Fahrenheit. Even when the heat wave relented, the earth was still feverishly warm. Today on windswept Ellesmere Island in the Canadian high Arctic, on a barren hillside overlooking ice-choked seas, fossil tree stumps mark the former site of an Eocene swamp forest once inhabited by flying lemurs, giant tortoises, hippolike animals, and alligators. Worst-case carbon dioxide emissions and climate sensitivity models promise to return our modern planet to this Eocene steam bath.

One proposed cause for the high-CO_2 hothouse that began in the age of dinosaurs and reigned in this early mammalian heyday is—once again—India. The subduction zones that dragged the island continent across the ocean, pulling it toward Asia, shoved the ocean floor down into the earth and devoured thousands of miles of carbonates laid down by dead sea life over the ages. Carbon dioxide from the consumed rock was continuously gassed out above, in a vanguard of volcanoes. When India crashed into Asia around 45 million years ago, this CO_2 factory—in operation for tens of millions of years—shuttered its doors and the volcanoes went quiet. As the collision pushed the Himalayas into the sky, these volcanic rocks and this newborn mountain range began to weather, drawing down CO_2 even further. As with the creation of the Appalachians and the Ordovician ice age 400 million years prior, when the uplift and weathering of the Himalayas began, the long, slow decline to the modern ice age was set in motion.

Eventually Antarctica, long a lush, forested preserve, began to separate from Australia, bringing an end to the last vestige of the supercontinent Gondwana. As the southernmost continent began to grow an ice cap and cooler and dryer climates spread across the globe, the Eocene ended with a chill 34 million years ago. This transition, from the long-standing greenhouse climate to a more modern climate with ice at the poles, caused a major turnover in animal life. Bizarre mammals like the knobby-headed, rhinoceros-like brontotheres vanished at this first blush of polar ice. Grasslands and savannas familiar to us today began to spread, taking over from primeval forests. This changeover is called "the Grande Coupure" (French for "big break"). But for the most part, extinctions and originations in the Cenozoic continued as they always do, with species living full natural lives before succumbing to the changing of geological seasons, blessedly free from the indiscriminate slaughter of mass extinctions. Unfairly passed over in the public imagination, the world since the age of dinosaurs has been a wild one, featuring everything from dinosaur-sized hornless rhinoceroses to godlike 60-foot megalodon sharks.

Then, a mere 3 million years ago, as carbon dioxide continued its faltering ebb and North and South America joined hands at Panama—a marriage that rerouted global ocean circulation—the top of the planet began to freeze over as well. The North Pole has probably remained mostly frozen ever since—that is, until our own time, when it's expected to melt away in the summers of the coming decades.

When the earth had cooled enough, about 2.6 million years ago, the wobble of the planet began to dominate the climate, tilting the earth in and out of the sunshine and thrusting the entire planet in and out of great ages of ice. When these periodic wobbles tilted the earth away from the sun in the summer, the ice could march across the continents in colossal sheets more than a mile thick. Winter came to the earth, clasping it in an icy embrace for tens of thousands of years. Over the past few million years, these wobbles in space and the regular changes to Earth's orbit have thrust the planet in and out of perhaps more than fifty cycles of advancing and retreating ice.

Which brings us to today. We find ourselves sandwiched between great ice ages, in a brief interglacial of warmth for a few thousand tenuous years, like the dozens of warm respites that have come and gone before. We should not expect this pleasant vacation to last much longer than it already has. In a geological moment, we should expect to be cast back into a great glacial age, during which New York City will look like the edge of Antarctica, with the Empire State Building an insignificant speck beside the icy face of the continental ice sheets. If the ice age returns, the seas will plummet 400 feet, pushing our familiar coastlines hundreds of miles offshore and connecting Australia to Asia, and Asia to North America. Later we'll return to this long-term forecast, which has been thrown into chaos by human intervention.

Curiously, the wild climate swings of the past few million years—in and out of punishing ice ages—have caused precious few extinctions. Unlike *Isotelus rex* or *Dunkleosteus*, which perished in earlier glaciations in earth history, woolly mammoths, giant ground sloths, enormous marsupials, and armadillos the size of cars seem to have survived the many swings between ice ages and warm interglacials of recent geological history with good humor, shifting their ranges to accommodate the finicky planet.

Then, a geological moment ago, the world lost half its enormous land mammals.

These are known as the "near-time" extinctions because, to geologists, events that happened only a few thousand years ago might as well have happened yesterday. These near-time extinctions, which represent the biggest hit to large land vertebrates since the biblical chaos at the end of the Cretaceous, follow a pattern unlike any other: avoiding the marine realm altogether, leaving the flora virtually intact, and primarily affecting large, charismatic land mammals.

After millions of years of relative stability, even through countless punishing climate swings, a strange wave of extinctions suddenly swept across the planet, eerily shadowing the heroic migrations of the recently evolved African primate species *Homo sapiens*. Starting only a few tens of thousands of years ago, the extinctions jumped from continent to continent, then to remote islands, and they continue unchecked to the modern day. The idea of man-made extinctions evokes images of gasoline-chugging chainsaws melting through old-growth timber, or industrial fishing trawlers sterilizing the seabed with rusty submarine plowshares, but in fact, humanity's gain has been biodiversity's loss since birth.

Sometime between 40,000 and 50,000 years ago, Australia lost its marsupial lions and its giant kangaroos, which were far larger (and slower) than those now living. It lost its diprotodons, giant lumbering herbivores the size of rhinoceroses—the largest marsupials to ever live. It lost its giant flightless birds, more than 6 feet tall. It lost a giant python, two species of terrestrial crocodiles, and an enormous monitor lizard called *megalania*, which, stretching some 15 feet, looked as though it got lost en route to the Triassic. It lost absolutely every animal on land weighing more than 100 kilograms (220 pounds). The wave of extinction struck, not during any unusual climate perturbation or asteroid impact, but at about the same time that the first humans arrived in Australia.

When modern humans first spread into Europe and Asia, the local fauna suffered a more prolonged period of extinctions—extinctions that claimed Eurasia's straight-tusked elephants, its woolly mammoths, its woolly rhinoceroses, and its not so woolly rhinoceroses, as well as its hippopotamuses, its giant deer (sporting the world's most flamboyant antlers), its cave bears, its cave lions, and its spotted hyenas. Eurasia also lost its Neanderthals—that other hominid that used tools and fire and buried their dead. The Neanderthal encounter with modern humans was devastatingly brief, though their genes live on in Europeans and Asians, whose love apparently transcended species.

The extinction of woolly mammoths, viewed hazily by the public as sometime back there with dinosaurs, is so recent that it's possible to eat woolly mammoth meat retrieved from the snow, as science writer Richard Stone witnessed one Russian comrade do on a trip to Siberia. "Even after a few shots of vodka, he said, 'It was awful. It tasted like meat left too long in the freezer.'" Spread across eastern Europe and Russia are the scattered remains of settlements featuring houses constructed entirely from mammoth bones, including the Ukraine's stunning Mezhyrich site, which includes the bones from some 150 animals.

Around 12,000 years ago, humans arrived in North America. At the same time, after millions of years of relative stability—again, even through wild shifts in climate—North America lost a staggering array of megafauna. The continent was home to a suite of animals far surpassing in grandeur that found on any modern African savanna. It lost its four species of mammoths, its elephant-like gomphotheres, and its giant ground sloths—some towering 15 feet tall on their hind legs. It lost its giant armadillos weighing more than a ton; beavers the size of bears; bears, like *arctodus*, that were far larger than any now living; and giant peccaries, tapirs, stag moose, capybaras, wild dogs, dwarf antelopes, brush oxen, woodland musk oxen, and mastodons.

Spores of a fungus that lived and relied on mastodon dung hint that this extinction wasn't due to natural forces like a shift in vegetation or climate change. The spores plummeted—indicating the disappearance of the mastodons and other megafauna they depended on—even as the animals' preferred spruce forests were spreading. Native American kill sites, as well as computer models simulating the relative ease of overhunting the megafauna to extinction in only a few generations, point to another culprit.

North America also lost its many camels, which originated and evolved on the continent, only later spreading out into Asia and Africa. When camels were experimentally employed in military convoys across the Southwest in the 1850s, Lieutenant Edward Beale—unaware of the animals' ancestral connection to the land—was pleasantly surprised by their uncommon effectiveness. Happily marching across their evolutionary homeland, they ate "the otherwise worthless weeds and other plants shunned by livestock, including creosote bush growing along the right of way in New Mexico."

North America lost its American zebras as well as its horses. The story of horses in North America is a curious one. Horses evolved on the continent over millions of years, then suddenly went extinct around 12,000 years ago, only to be reintroduced a few thousand years later by Spanish colonists. If they persist on the continent for millions of years from now, geologists of the far future probably won't detect this strange millennia-long absence.

Unable to scavenge the previously plentiful carcasses of North American megafauna, the continent lost its teratorns—among the largest birds ever to fly—along with many of its condors. It lost its dire wolves and its saber-toothed cats. It lost an American cheetah, as well as one of the largest cats ever, the American lion—bigger even than its African cousin. You can find the remains of many of these animals where they died; in downtown Los Angeles, for instance, their bones are preserved in the muck of natural asphalt in the tar pits of La Brea, along the city's congested Miracle Mile.

All of these animals roamed North America until so recently that, to future geologists, they will appear to have gone extinct essentially right now. That our age is thought to be less epic than those worlds on offer at natural history museums is only an illusion. Our newly denuded landscapes are howlingly barren and impoverished from only a geological moment ago.

But the menagerie lives on in evolutionary ghosts. In North America, the fleet-footed pronghorns of the American West run laughably faster than any of their existing predators. But then, their speed isn't meant for existing predators. It might be a vestige of their need to escape constant, harrowing pursuits by American cheetahs—until a geological moment ago. The absence was palpable to me as I rode a train past New Mexico's Kiowa National Grassland, an American Serengeti, windswept and empty except for a lone wandering pronghorn still running from ghosts.

Other evolutionary shadows of the Pleistocene live on in the produce aisle. Seeds in fruit are designed to be eaten and dispersed by animals, but for the avocado this makes little sense. Their billiard ball–sized cores, if swallowed whole, would at the very least make for an agonizing few days of digestive transit. But the fruit makes a little more sense in a land populated by tree-foraging giants, like the sometimes dinosaur-proportioned ground sloths, who swallowed the seeds and hardly noticed them. The ground sloths disappeared a geological moment ago, but their curious fruit, the avocado, remains.

The extinction of the ground sloths and the rest of the American megafauna is so recent that there remain to this day caves in the Grand Canyon filled with giant ground sloth shit. In likely the most moving passage ever written about wading through shit, the late University of Arizona paleontologist Paul Martin described an expedition to the Grand Canyon's Rampart Cave:

> Slowly proceeding deeper into the cave, we fell silent as in a cathedral. . . . In single file we walked into a trench,

> through sloth dung. When we stopped we stood chest deep in layers of stratified sloth dung. There was no perceptible airflow, but the deposit had lost any trace of ammonia or other odors of decaying manure; the air smelled resinous, like incense. No one spoke a word. In the stillness I felt the hair rise on the back of my neck. One did not need to be a Sufi or a mystic to sense that this dimly lit, low-ceilinged chamber was a sacred sanctuary. More than a sepulcher for the dead, Rampart Cave venerated the extinct.

No one drew more fire for the unpopular idea that indigenous populations were responsible for these extinctions than the late Martin, who first proposed his "overkill" theory in the 1960s. Many postmodern social scientists and anthropologists were repelled by the idea that First Peoples, already dehumanized and decimated by colonialism, could be responsible for staggered, global waves of extinction. One of Martin's most vocal critics was his colleague at the University of Arizona, political science professor Vine Deloria Jr., who eventually espoused a version of Native American creationism in which Native Americans originated in North America and had always been there. Overwhelming scientific evidence to the contrary from genetics, archaeology, and paleontology, pointing to an arrival from Asia roughly 12,000 years ago and a subsequent devastation of North American megafauna, were viewed as further acts of Western cultural imperialism. But Martin made every effort to note that it would be ridiculous to hold prehistoric people responsible for running afoul of modern conservation ideals, holding that, "should we, in the next 12,000 years, cause as few extinctions of large mammals as the Native Americans have in the 12,000 calendar years since the days of the ground sloths, we would be able to consider ourselves incredibly lucky." Near the end of his life, Martin even advocated restoring our ecologically impoverished landscape by importing elephants and camels from Africa and Asia to repopulate the American West.

Among his more scientifically literate colleagues, critics of the overkill theory pointed to climate change at the end of the last ice age as an alternative explanation for the extinctions, exonerating these early pioneering humans. And North America did indeed experience dramatic climate changes during the transition out of the most recent ice age. But it had done so countless times before during the Pleistocene, and the climate changes at the end of the most recent ice age were certainly no larger or more intense than the many previous oscillations between glacial and interglacial times. The beasts of the Pleistocene had easily navigated the earlier changes, shifting their ranges to track their preferred habitat. The changing climate might have added an extra element of instability, making the biosphere more vulnerable to the sort of disruption imposed by multiplying bands of skilled hunters transforming the landscape with fire as they spread. But there's no reason to think that the megafauna of North America would have gone extinct without the introduction of humans—the ultimate invasive species. It's also nearly impossible to appeal to climate change to explain why nocturnal animals tended to fare better in the extinctions. The same can be said for plants, few of which went extinct. The same ground sloth dung that Martin encountered in the Grand Canyon revealed a diet of plants that still thrive in the arid landscapes of North America and are happily eaten by bighorn sheep and wild burros. It seems unlikely that the slow, defenseless giant ground sloths disappeared for want of food.

Finally, there existed control groups to test Martin's theories. On islands and landmasses that remained undiscovered by humans for thousands of years, megafauna survived the climate changes at the end of the Pleistocene, as they had many times before, only to be destroyed when humans eventually arrived on their shores.

The last ground sloths might have vanished from mainland North America 10,000 years ago, but in 2005 Martin's former student, University of Florida paleontologist David Steadman, found fossils of a species that lingered on in Hispaniola and Cuba for an additional 5,000 years. When the West Indies were first settled by humans, these ground sloths of the Caribbean quickly disappeared as well.

Amazingly, woolly mammoths also persisted unseen on extremely remote islands, even as late as the golden age of Egyptian pyramid construction. The anachronistic mammoths found themselves marooned, but safe, on Wrangel Island off Siberia, and on Saint Paul in the godforsaken Pribilof Islands in the Bering Sea, far north of the Aleutian Islands. These refuges went undiscovered by humans and the mammoths survived, while their mainland relatives were felled by extinction thousands of years before.

Similarly, Steller's sea cow, a gigantic *30-foot-long* manatee cousin, was eradicated from the North Pacific coast around 12,000 years ago but managed to survive unmolested as a small remnant population on the uninhabited Commander Islands off Russia until the eighteenth century. The Commander Islands were discovered in 1741 by fur traders. The *12-ton* giant was hunted to extinction on this, its final redoubt—and thus its entire world—within three decades of human discovery.

While the damage was done on the continents over 10,000 years ago, islands continued to suffer wave after wave of extinction over the centuries as they were discovered by ancient, rugged explorers. About 2,000 years ago, after Indonesians made a remarkable journey to Madagascar over the Indian Ocean in outrigger canoes, they came ashore and razed the local fauna. The pulse of extinction claimed an aardvark relative and seventeen species of lemurs, the largest of which, *archaeoindris,* was the size of a gorilla. Madagascar also lost its hippopotamuses, its giant tortoises, and its staggering elephant birds, which stood more than 10 feet tall and laid eggs that, with a capacity of well over 2 gallons, were the largest eggs known of any animal ever, even (nonbird) dinosaurs. These enormous eggshells are not difficult to find on the island, where they "litter the ground like the wrack of clamshells." They must have provided feasts for the early Malagasy.

In the last few hundred years, as brave Polynesians set out upon the Pacific to miraculously colonize tiny atolls and archipelagoes separated by thousands of miles—from New Caledonia to Hawaii to Easter Island to the Pitcairns—local island fauna, including thousands of species of flightless birds, along with countless land snails and other animals, were eradicated. But hunting isn't the only weapon in humanity's armamentarium of extinction. This island fauna might have been destroyed primarily by our furry cargo, like rats and pigs.

In New Zealand, the fossil record for outlandish flightless birds called moas shows the colossal birds—some taller than a basketball hoop—handling the finicky climate of the Pleistocene sanguinely, marching up and down the island as the earth wobbled in and out of the sunshine. But 500 years ago, the Maori arrived in New Zealand and the moas disappeared. The extinction puzzled UCLA ornithologist and geographer Jared Diamond, also the author of *Guns, Germs, and Steel.* He thought the idea that human artifice alone could cause the extinctions was ridiculous on its face. He shed his skepticism while working in the remote and inaccessible Gauttier Mountains of New Guinea, where he encountered an utterly fearless tree kangaroo.

> Until I had worked in the Gauttiers, I was mystified to understand how the few Maoris in the vastness of New Zealand's South Island could have killed all the moas, and how

> anyone could take seriously the Mosimann-Martin hypothesis of Clovis hunters eliminating most large mammals from North and South America in a millennium or so. I no longer find this at all surprising when I recall the large kangaroo *Dendrolagus matschiei* remaining on a tree trunk at a height of 2 meters, watching my field assistant and me as we talked nearby in full sight.

It is this naïveté about humans that may account for many of the extinctions. Besides, how could any animal have any idea that this strange, bipedal mammal, smaller than some deer and lacking fearsome claws or teeth, could be so lethal? In fact, by the twentieth century, every land animal vulnerable to this sort of ignorance had already paid dearly. But uninhabited and undiscovered, Antarctica avoided the waves of extinctions that struck every other continent. It wasn't until Victorian-era explorers arrived on its shores that humans encountered the sort of nutritious and fearlessly friendly fauna that was likely to have welcomed First Peoples to new landmasses over the previous 50,000 years. Upon arriving in 1912, Norwegian explorer Roald Amundsen couldn't believe his good fortune. "We live in a veritable Never-Never land," he wrote about the new continent. "Seals come up to the ship and penguins to the tent, and allow themselves to be shot." The unfamiliar animals hadn't had time to develop what Darwin called a "salutary dread" of man. To the Native Americans, Indigenous Eurasians, and Aboriginal Australians—presumably virtuosic hunters all—their new homelands must have similarly seemed a lavish never-never land. Encountering huge aggregations of fearless prey, huddled around watering holes, must have proved an irresistible bounty.

That said, the relative intactness of African megafauna, which spent the most time in the company of humans, has been cited as evidence against the overkill hypothesis. But it might be the exception that proves the rule. Slowly coevolving with people over 2 million years as hominids became ever more adept at wielding technology and strategy in pursuit of prey, these animals, alone among their global counterparts, had the requisite evolutionary time and harrowing experience to learn their "salutary dread of man." Still, even Africa lost 21 percent of its megafauna, with larger animals getting hit the hardest.

British geologist Anthony Hallam (with a somewhat unseemly triumphalism) cites this record of precolonial ecological ruin to "dispel once and for all the romantic idea of the superior ecological wisdom of non-western and pre-colonial societies. The notion of the noble savage living in harmony with Nature should be dispatched to the realm of mythology where it belongs. Human beings have never lived in harmony with nature."

That the human project since its birth, and human flourishing in general, seems to have played out at the expense of the rest of the natural world is one of the stark and unsettling discoveries of science.

This destructive human shadow has grown in the past few centuries, and the list of extinctions in the very near past is tragic and well known: from Australia's marsupial Tasmanian tiger to North America's passenger pigeons (both of which spent their last days in zoos) to Europe's great auks and Mauritius's dodo. In China, dams, fishing gear, and boat traffic have driven the *baiji,* an almost blind river dolphin, to extinction in only the last decade. And in 2015 the image of the very last male northern white rhino, being guarded by armed Sudanese wildlife rangers at the end of its million-year run on the planet, made headlines around the globe. Countless other lost species will never be known, plowed to oblivion by trawlers on the ravaged continental shelves of the ocean, or lost in smoldering tracts of cleared rainforest.

So what innovation had evolution stumbled upon with human beings? What could account for this much destruction, this fast, rendered entirely by a single species of primate? If deep roots, thick woody tissues, and seeds in early land plants did in life during the Late Devonian, what could account for the nearly instantaneous dispersal of *Homo sapiens* across the globe and their subsequent domination of the natural environment?

Culture might have something to do with it.

By culture, of course, I'm not referring to Monet's *Water Lilies* or the plays of August Wilson, but rather the ability of *Homo sapiens* to transmit information from generation to generation, not only through our genetic code, like the rest of the animal kingdom, but through language, behavior, and technologies like writing. It is culture that enables us to adapt to the environment as it changes on the fly, rather than being compelled to wait around for the hammerblows of natural selection to painfully correct us.

Culture, like DNA, is information. As such, it propagates and evolves based on its effectiveness at transmitting itself. Like genes, information encoded in language or behavior that keeps people alive or confers some material advantage on them is good at getting itself disseminated. This could include information about things like crop rotation, or how to build boats, weapons, or clothing. And as Tufts philosopher Daniel Dennett has claimed, this process need not ever involve human ingenuity. The design of Polynesian boats, Dennett says, was sculpted by something like evolution by natural selection. Poor boat designs—ones whose occupants didn't return to port—were not adopted by successive generations of boat builders. Instead, boat builders adopted only those designs, selected by the sea, that survived the journey.

But even if the design of the boat was conceived for reasons unavailable to its makers, shaped by the sea rather than by an all-wise designer, the steady accumulation of improvements over successive generations of boats, through cultural evolution, leaves—in almost no geological time—an amazing vessel capable of overcoming natural barriers even as daunting as the Pacific Ocean. This ability to transmit mutable information down through the generations about superior boat-making techniques (or hunting methods, or ways to make clothes with animal pelts, or metallurgical expertise) allowed technology to accumulate countless new modifications and tweaks and thus to become ever more adaptive in an eyeblink of evolutionary time. The invention of the written word allowed this information about manipulating the physical world—that now resides and mutates outside the genome, in books, magazines, newspapers, science journals, and, most recently, the Internet—to be dispersed ever more widely. There is a straight line of cultural evolution—a cultural clade—from spears to nuclear weapons. Culture has allowed us to cast off the shackles of evolutionary time.

Today these tens of thousands of years of cultural evolution have given us a world where we have gained such mastery over the physical environment that we hold the knobs of the entire earth system—and are twisting them violently.

One innovation in particular has turned us into a truly geological force: our global effort to take as much ancient carbon from the rock record as possible and ignite it all at once in the atmosphere. This is a superpower normally reserved for continental flood basalts.

For hundreds of millions of years, the planet has been burying these huge stores of carbon in jungles of coal, or in blizzards of plankton at the bottom of the ocean. In just a few centuries, humanity is trying to light a match to all of it. In many ways this geological bonfire seems grotesque and unnatural, but when viewed in light of earth history, it looks like one of those major metabolic innovations that happen every few hundred million, or billion, years. Throughout history, biology has continuously invented new ways of more efficiently using untapped reservoirs of energy, ultimately derived from sunlight hitting the earth. One way to capture this solar power is photosynthesis in plants. Another way to capture it is to eat the plants that store that solar power in their leaves as sugars. Still another way is to eat and digest the mouse that eats those plants, shunting the solar power even higher up the food chain. But at root, it's all about capturing the energy from photons streaming off a star exploding 93 million miles away in outer space. Burning the carbon from ancient plants in coal and gasoline to power complex, energy-intensive societies is just the latest such biological innovation.

"Coal is a resource that for 300 million years no one figured out how to use," said Stanford's Jonathan Payne. "It was just sitting there. It was an energy reservoir, and we figured out how to use it."

As a result of this innovation, human civilization is now propped up by a continuous explosion of energy, a global mega-metabolism, with hundreds of millions of years' worth of sunlight being released all at once in combustion engines and power plants. Carbon dioxide is a by-product of this new civilizational metabolism, and we now emit 100 times more CO_2 each year than volcanoes. This far outstrips the ability of the earth's thermostat to keep up through rock weathering and ocean circulation, operating as those processes do on 1,000- to 100,000-year timescales.

But the carbon cycle is not the only earth system getting short-circuited thanks to human ingenuity. We're also living through the largest disruption to the earth's nitrogen cycle in 2.5 billion years. This might sound like arcane geochemistry, but the ramifications are extraordinary. Plants need nitrogen to live. It's what Miracle-

Gro is made of. Until the twentieth century, almost all biologically available nitrogen was fixed by microbes in the roots of legumes. Now humans synthesize this fertilizer from fossil fuels, fixing twice as much nitrogen as the natural world does every year. Before the twentieth century, the human population was limited in size by the number of crops it could grow, which was limited by the amount of nitrogen fertilizer available in nature from sources like manure. In 1909 the German chemist Fritz Haber invented the process of artificial nitrogen fixation and destroyed these natural limits.

The subsequent explosion in agriculture is directly responsible for the existence of the billions of people alive today who would otherwise not be here, most likely including myself. This is what accounts for the unnerving right angle on the graph of human population in the twentieth century. It took 200,000 years, until 1850 or so, for the global population to grow to one billion people. Now we're adding an extra billion people to the population every decade or so, born aloft on this artificial surfeit of plant food.

Along with all of those people, artificial nitrogen fixation is also responsible for huge tracts of dead zones in the oceans around the world, as fertilizer runoff from industrial agriculture causes Devonian/Permian/Triassic-style blooms of phytoplankton that rob the ocean of oxygen. And this huge perturbation to the nitrogen cycle feeds back on the carbon cycle as those billions of new people demand ever more ancient carbon from the rocks to burn to subsidize their modern lifestyles. If the man-made extinctions at the end of the last ice age had much to do with how humanity moved through the world, the extinctions of today have more to do with how the world moves through us, with the life-sustaining cycles of nitrogen and carbon being rerouted and warped by the human project.

But besides swelling the population of our subspecies of primate and creating metastasizing dead zones in the ocean, all that extra plant food has created a still crazier inversion of the earth's fauna.

It stands to reason that, until very recently, all vertebrate life on the planet was wildlife. But astoundingly, today wildlife accounts for only 3 percent of earth's land animals; human beings, our livestock, and our pets take up the remaining 97 percent of the biomass. This Frankenstein biosphere is due both to the explosion of industrial agriculture and to a hollowing out of wildlife itself, which has decreased in abundance by as much as 50 percent since 1970. This cull is from both direct hunting and global-scale habitat destruction: almost half of the earth's land has been converted to farmland.

The oceans have endured a similar transformation in only the past few decades as the industrial might developed during World War II has been trained on the seas. Each year fishing trawlers plow an area of seafloor twice the size of the continental United States, obliterating the benthos. Gardens of corals and sponges hosting colorful sea life are reduced to furrowed, lifeless plains. What these trawlers have to show for all this destruction is the removal of up to 90 percent of all large ocean predators since 1950, including familiar staples of the dinner plate like cod, halibut, grouper, tuna, swordfish, marlin, and sharks. As just one slice of that devastation, 270,000 sharks are killed every single *day*, mostly for their tasteless fins, which end up as status symbol garnishes in the bowls of Chinese corporate power lunches. And today, even as fishing pressure is escalating, even as the number of fishing boats increases, even as industrial trawlers abandon their exhausted traditional fishing grounds to chase down ever more remote fish stocks with ever more sophisticated fish-finding technology, global fish catch is flatlining.

Closer to shore, coral reefs, the wellspring of the ocean's biodiversity, have declined in extent by as much as a third since only the 1980s. These paradises are plagued by overfishing, pollution, and invaders, but 500 million people, many of them poor and living in the developing world, rely on them for food, storm protection, and jobs. As in the handful of reef collapses of the geological past, modern reefs are expected to collapse from warming and ocean acidification by the end of the century, and possibly much sooner. During record-breaking temperatures in 1997–1998, 15 percent of the world's reefs died. And in 2015, amid strangely tepid water, a wave of death once again swept through the already battered reefs of southern Florida and the Keys, wiping out huge tracts of coral, some of which had survived for hundreds of years. Meanwhile in Hawaii, the AP reported, "the worst coral bleaching the islands have ever seen" was under way. The die-offs are part of an expected global bleaching event like 1997–1998 that will sweep across the Pacific, fueled by warming seas. And it won't be the last.

As we've seen in previous chapters, some kinds of plankton, like the fluttering pteropods that make up to 50 percent of the diet for fish like Pacific salmon and are a foundational part of the Antarctic ecosystem, are already dissolving in the Pacific Northwest and around Antarctica. They could be wiped out entirely in the Southern Ocean by 2050. And as sea ice declines, so too will krill, which feed on the algae gardens on the underside of the ice. Krill is also sensitive to acidifying oceans, and scientists expect up to a 70 percent decline in Antarctic krill by the end of the century from ocean acidification alone. Krill feeds seals, penguins, and whales, but is being replaced in the ecosystem by gelatinous colonial tubes called salps. Where krill converts solar power from plankton into whales, salps have little nutritional value and few predators. A Southern Ocean without pteropods or krill is an utterly ravaged one.

And the demands we make of the oceans in the coming decades will only escalate as the population grows to perhaps more than 11 billion people—with most of that growth coming from the poor, developing world, which relies disproportionately on seafood from doomed coral reefs.

So things don't look so good, no matter where we look. Yes, the victims in the animal world include scary apex predators that pose obvious threats to humans, like lions, whose numbers have dropped from 1 million at the time of Jesus to 450,000 in the 1940s to 20,000 today—a decline of 98 percent. But also included have been unexpected victims, like butterflies and moths, which have declined in abundance by 35 percent since the 1970s.

Like all extinction events, so far this one has been phased and complex, spanning tens of thousands of years and starting when our kind left Africa. Other mass extinctions have similarly played out over tens of thousands, hundreds of thousands, or even, as in the Late Devonian, millions of years. To future geologists, then, the huge wave of extinctions a few thousand years ago as First Peoples spread out into new continents and remote archipelagoes will be all but indistinguishable from the current wave of destruction loosed by modernity and its growing appetites.

Now for the crazy part—and the part that should shed some light on how bad the Big Five mass extinctions actually were. Despite this record of devastation, and despite the casual gloom of the many science journalists and conservation nonprofits peddling the reality of a current sixth mass extinction on par with the first five, humanity has not yet come anywhere even remotely close to death tolls of the major mass extinctions of the past half-billion years . . . yet. In the past 400 years, there have been documented extinctions of some 800 species. This is a tragedy, to be sure, and likely a massive undercount, but when divided by the 1.9 million species known, 800 extinct species amounts to an extinction of less than one-tenth of 1 percent—a far cry from the End-Permian, during which, generously rounding up, almost 100 percent of complex life on earth was killed.

Fish might have been decimated by industrial-scale fishing in the past few decades, but very few have gone extinct: each year sperm whales eat as much seafood as we do—though a fraction of their historic population—there are still hundreds of thousands of sperm whales. There has not been anything like the total collapse of life on earth seen at the end of the Permian, or at any of the other mass extinctions, either on land or at sea. In fact, biodiversity is still flourishing. Look out your window and you may see a verdant place, filled with birdsong and fattening squirrels. Even with the loss of all those giant ground sloths and mammoths and mastodons and dodos and rhinos and tree frogs and passenger pigeons and pangolins and baiji, in the big picture we've landed only a body blow on this glorious biosphere. Especially compared to the global holocausts of deep time.

"The headlines are not just inaccurate," writes futurist Stewart Brand on the premature obituary for the planet that has become de rigueur in some circles. "As they accumulate, they frame our whole relationship with nature as one of unremitting tragedy. The core of tragedy is that it cannot be fixed, and that is a formula for hopelessness and inaction. Lazy romanticism about impending doom becomes the default view."

In fact, from a geological perspective, the planet may be more resistant to mass extinctions today than at any point in its history. For one thing, we don't find ourselves with the carbon-jamming geometry of a Pangaean supercontinent (though humans, by introducing invasive species around the world, have re-created some of the negative aspects of supercontinent living), nor are we in the stranded island world of the Ordovician, where escape routes were foreclosed (though habitat fragmentation may pose a similar challenge). But perhaps the most important aspect of the modern earth's resilience is the change that has taken place over the past few hundred million years in the oceans, which are as oxygenated as they've ever been. Some of earth's most salutary changes might be thanks to its most modest inhabitants: plankton.

Plankton has gotten bulkier and heavier over time—with huge implications for the ocean and for life on earth. Today's single-celled, armored drifters—creatures like foraminifera, and the more plant-like diatoms, and coccolithophores—are microscopic to our eyes but are still enormous compared to the single-celled plankton of the Paleozoic, a suite that was dominated by bacteria and green algae. This sort of modern plankton is also freighted with mineral ballast, an extra load that, combined with its size, allows it to sink much further into the deep ocean before being consumed again by life. This has huge consequences, as consumption of this biological snowfall in the oceans uses up oxygen. If plankton can sink deeper into the ocean before being consumed, then so too will the ocean's Oxygen Minimum Zone (OMZ), the layer at which dissolved oxygen in the ocean is at its lowest. Today the OMZ is at about 600 meters down. But in the earth's past—when plankton particles were tinier and sank more slowly—the OMZ might have been much, much shallower, with devastating consequences for life. Today the OMZ is safely out of the range of the shallow continental shelves, where most sea life lives. But in the Paleozoic, when the shallower OMZ began to rise (from sea level rise, global warming, or nutrient pollution, for instance), it spilled onto the continental shelves, bringing oxygen-starved waters up to the shallows where it could smother sea life. The result was mass extinction.

"In the Paleozoic we don't really even comment on ocean anoxic events, they're so common," said Jonathan Payne. "In the Mesozoic we comment on them as these interesting things, but by the Cenozoic we basically just don't even find them."

What is going on today is extremely unusual. We're hunting and destroying animals at unfathomable rates, but if humanity were to disappear tomorrow, the planet might quickly recover. If we stopped dumping carbon into the atmosphere and ocean, in a few thousand years it would come out of the system as limestone. But we're not likely to stop anytime soon, and alas, our plunder can't go on forever without unleashing geologically significant devastation.

In 2011, UC Berkeley paleontologist Anthony Barnosky and his colleagues published the paper "Has the Earth's Sixth Mass Extinction Already Arrived?" Judging by the coverage of the paper in the popular press, the answer seems to be an unequivocal yes. But in fact the paper predicted that the planet would reach Big Five levels of extinction only after hundreds to several thousands of years of continued and unabated environmental destruction. This outcome would still be instantaneous from a geological point of view, but from a human perspective the sixth mass extinction is still, thankfully, a little ways off.

Nevertheless, the paper did note that, like the projections of ice sheet collapse, this forecast might overlook some unexpected surprises. "Ecosystems may respond in a nonlinear fashion," Barnosky and his colleagues warned, "to gradual accumulation of environmental perturbations." According to these researchers, we might not even much notice the damage—that is, until we reach an "ecological threshold." At that point we could be in store for "large and sudden biotic changes."

In other words, there could be a tipping point.

At the 2014 annual meeting of the Geological Society of America, Smithsonian paleontologist Doug Erwin took the podium to address a ballroom full of geologists on the dynamics of mass extinctions and power grid failures—which, he claimed, unfold in the same way.

"These are images from the NOAA website of the US blackout in 2003," he said, pulling up a nighttime satellite picture of the glowing northeastern megalopolis, megawatts afire under the cold dark of space. "This is twenty hours before the blackout. You can see Long Island and New York City."

"And this is seven hours into the blackout," he said, pulling up a new map, cloaked in darkness. "New York City is almost dark. The blackout extended all the way up into Toronto, all the way out to Michigan and Ohio. It covered a huge section of both Canada and the United States. And it was largely due to a software bug in a control room in Ohio."

Erwin proposed that mass extinctions might unfold like these power grid failures: most of the losses may come, not from the initial shock—software glitches in the case of power grid failures, and asteroids and volcanoes in the case of mass extinctions—but from the secondary cascade of failures that follow. These are devastating chain reactions that no one understands. Erwin thinks that most mass extinctions ultimately resulted, not from external shocks, but from the internal dynamics of food webs that faltered and failed catastrophically in unexpected ways, just as the darkening eastern seaboard did in 2003. In the few hours of the blackout, the Northeast lost 80 percent of its power load—all because of an insignificant local hiccup.

"Because it was not clear how to manage that collapse—although after the fact it was clear that it should have been easily contained—it cascaded into failure of grids across the northeastern United States. . . . I mention this because it turns out that, from a mathematical point of view, the problem of understanding these food webs is exactly the [same] problem as understanding the nature of the power grid.

"There's a very rapid collapse of the ecosystem during these mass extinctions," he said.

I had written to Erwin to get his take on the now-fashionable idea that there is currently a sixth mass extinction under way on our planet on par with the Big Five. Many popular science articles take this as a given, and indeed, there's something emotionally satisfying about the idea that humans' hubris and shortsightedness are so profound that we're bringing down the whole planet with us.

Erwin thinks it's junk science.

"Many of those making facile comparisons between the current situation and past mass extinctions don't have a clue about the difference in the nature of the data, much less how truly awful the mass extinctions recorded in the marine fossil record actually were," he wrote me in an email. "I am not claiming that humans haven't done great damage to marine and terrestrial extinctions, nor that many extinctions have not occurred and more will certainly occur in the near future. But I do think that as scientists we have a responsibility to be accurate about such comparisons."

I had a chance to sit down with Erwin after his talk at the annual geology conference. My first question—about a rumor I had heard from one of his colleagues that Erwin had served as a sort of mass extinction consultant to Cormac McCarthy while the notoriously secretive author was constructing the post-apocalyptic world of *The Road*—Erwin coyly evaded. But on the speculative sixth mass extinction, he was more forthcoming.

"If we're really in a mass extinction—if we're in the [End-Permian]—go get a case of scotch," he said.

If his power grid analogy is correct, then trying to stop a mass extinction after it's started would be a little like calling for a building's preservation while it's imploding.

"People who claim we're in the sixth mass extinction don't understand enough about mass extinctions to understand the logical flaw in their argument," he said. "To a certain extent they're claiming it as a way of frightening people into action, when in fact, if it's actually true we're in a sixth mass extinction, then there's no point in conservation biology."

This is because by the time a mass extinction starts, the world would already be over.

"So if we really are in the middle of a mass extinction," I started, "it wouldn't be a matter of saving tigers and elephants—"

"Right, you probably have to worry about saving coyotes and rats.

"It's a network collapse problem," he said. "Just like power grids. Network dynamics research has been getting a ton of money from DARPA [Defense Advanced Research Projects Agency]. They're all physicists studying it, who don't care about power grids or ecosystems, they care about math. So the secret about power grids is that nobody actually knows how they work. And it's exactly the same problem you have in ecosystems.

"I think that if we keep things up long enough, we'll get to a mass extinction, but we're not in a mass extinction yet, and I think that's an optimistic discovery because that means we actually have time to avoid Armageddon," he said.

Erwin's other point, that the magnitude of the Big Five dwarfs humanity's destruction thus far, is a subtle one. He's not trying to downplay the tremendous destruction wrought by humans, but reminding us that claims about mass extinctions are inevitably claims about paleontology and the fossil record.

"So there are estimates of what the standing crop of passenger pigeons was in the nineteenth century," said Erwin. "It's like 5 billion. They would black out the sky."

Passenger pigeons all but serve as the mascot of the "sixth mass extinction," their extirpation an ecological tragedy on a massive scale, and proof that humans are a geologically destructive force to be reckoned with.

"So then you ask: in a non-archaeological context, how many fossil passenger pigeons are there? How many records are there of fossil passenger pigeons?"

"Not many?" I offered.

"Two," he said.

"So here's an incredibly abundant bird that we wiped out. But if you look in the fossil record, you wouldn't even know that they were there."

Erwin likes to recall a talk he once went to by an ecologist who had documented the troubling losses he had seen over his career in high-altitude rainforests.

"He was using this as an example of the destruction of plants in these cloud forests in Venezuela, all of which could be completely true," Erwin said. "The problem is, the probability of finding one of those cloud forests in the fossil record is zero."

The fossil record is incredibly incomplete. One rough estimate holds that we've only ever found a tantalizing 0.01 percent of all the species that have ever existed. Most of the animals in the fossil record are marine invertebrates, like brachiopods and bivalves, of the sort that are both geologically widespread and durably skeletonized. In fact, though this book (for narrative purposes) has mostly focused on the charismatic animals taken out by mass extinctions, the only reason we *know* about mass extinctions in the first place is from the record of this incredibly abundant, durable, and diverse world of marine invertebrates, not the big, charismatic, and rare stuff like dinosaurs.

"So you can ask, 'Okay, well, how many geographically widespread, abundant, durably skeletonized marine taxa have gone extinct thus far?' And the answer is, pretty close to zero," Erwin pointed out. "It's not that we haven't lost a lot of stuff. The problem is that we could have lost taxa of the same kind as we've lost now without ever seeing it in the fossil record."

When mass extinctions hit, they don't just take out big charismatic megafauna, like elephants, or niche ecosystems, like cloud forests. They take out hardy and ubiquitous organisms as well—things like clams and plants and insects. This is incredibly hard to do. But once you go over the edge and flip into mass extinction mode, nothing is safe. Mass extinctions kill almost everything on the planet.

While Erwin's argument that a mass extinction is not yet under way might seem to get humanity off the hook—an invitation to plunder the earth further, since it can seemingly take the beating (the planet has certainly seen worse)—it's actually a subtler and possibly far scarier argument.

This is where the ecosystem's nonlinear responses, or tipping points, come in. Inching up to mass extinction might be a little like inching up to the event horizon of a black hole—once you go over a certain line, a line that perhaps doesn't even appear all that remarkable, all is lost.

"So," I said, "it might be that we sort of bump along where everything seems okay and then . . ."

"Yeah, everything's fine until it's not," said Erwin. "And then everything goes to hell."

Or put another way, mass extinctions may unfold the same way that a dissolute character in Hemingway's *The Sun Also Rises* explains that bankruptcies do: "Two ways. Gradually and then suddenly."

"The only hope we have in the future," Erwin said, "is if we're not in a mass extinction event."

8

THE NEAR FUTURE

The earth is fast becoming an unfit home for its noblest inhabitant, and another era of equal human crime and human improvidence, and of like duration with that through which traces of that crime and that improvidence extend, would reduce it to such a condition of impoverished productiveness, of shattered surface, of climatic excess, as to threaten the depravation, barbarism, and perhaps even extinction of the species.
—*George Perkins Marsh, 1863*

Many of us share some dim apprehension that the world is flying out of control, that the center cannot hold. Raging wildfires, once-in-1,000-year storms, and lethal heat waves have become fixtures of the evening news—and all this after the planet has warmed by less than 1 degree Celsius above preindustrial temperatures. But here's where it gets really scary.

If humanity burns through all its fossil fuel reserves, there is the potential to warm the planet by as much as 18 degrees Celsius and raise sea levels by hundreds of feet. This is a warming spike of an even greater magnitude than that so far measured for the End-Permian mass extinction. If the worst-case scenarios come to pass, today's modestly menacing ocean-climate system will seem quaint. Even warming to one-fourth of that amount would create a planet that would have nothing to do with the one on which humans evolved, or on which civilization has been built. The last time it was 4 degrees warmer there was no ice at either pole and sea level was 260 feet higher than it is today.

I met University of New Hampshire paleoclimatologist Matthew Huber at a diner near campus in Durham, New Hampshire. Huber has spent a sizable portion of his research career studying the hothouse of the early mammals, and he thinks that in the coming centuries we might be headed back to the Eocene climate of 50 million years ago, when there were Alaskan palm trees and alligators splashed in the Arctic Circle.

"The modern world will be much more of a killing field than the PETM was," he said. "Habitat fragmentation today will make it much more difficult to migrate. But if we limit it below 10 degrees of warming, at least you don't have widespread heat death."

In 2010, Huber and coauthor Steven Sherwood published one of the most ominous science papers in recent memory: "An Adaptability Limit to Climate Change Due to Heat Stress."

"Lizards will be fine, birds will be fine," Huber said, noting that life has thrived in hotter climates than even the most catastrophic projections for anthropogenic global warming. This is one reason to suspect that the collapse of civilization might come long before we reach a proper biological mass extinction. Life has endured conditions that would be unthinkable for a highly networked global society partitioned by political borders. Of course, we're understandably concerned about the fate of civilization, and Huber says that, mass extinction or not, it's our tenuous reliance on an aging and inadequate infrastructure—perhaps, most ominously, on power grids—coupled with the limits of human physiology that may well bring down our world.

In 1977, when power went out for only one summer day in New York, whole swaths of the city devolved into something like Hobbes's man in a state of nature. Riots swept across the city, thousands of businesses were destroyed by looters, and arsonists lit more than a thousand fires. In 2012, when the monsoon failed in India (as it's expected to do in a warmer world), 670 million people—that is, 10 percent of the global population—lost access to power when the grid was crippled by unusually high demand from farmers struggling to irrigate their fields, while the high temperatures sent many Indians seeking kilowatt-chugging air-conditioning.

"The problem is that humans can't even handle a hot week to-day without the power grid failing on a regular basis," he said, noting that the aging patchwork power grid in the United States is built with components that are allowed to languish for more than a century before being replaced. "What makes people think it's going to be any better when the average summer temperature will be what, today, is the hottest week of the year in a five-year pe-riod, and the *hottest* temperatures will be in the range that no one has ever experienced before in the United States? That's 2050."

By the year 2050, according to a 2014 MIT study, there will also be 5 *billion* people living in water-stressed areas.

"Thirty to fifty years from now, more or less, the water wars are going to start," Huber said.

In their book *Dire Predictions,* Penn State's Lee Kump and Mi-chael Mann describe just one local example of how drought, sea level rise, and overpopulation may combine to pop the rivets of civilization:

> Increasingly severe drought in West Africa will generate a mass migration from the highly populous interior of Ni-geria to its coastal mega-city, Lagos. Already threatened by

> rising sea levels, Lagos will be unable to accommodate this massive influx of people. Squabbling over the dwindling oil reserves in the Niger River Delta combined with potential for state corruption will add to the factors contributing to massive social unrest.

"Massive social unrest" here being, of course, a rather bloodless phrase masking the utter chaos coming to a country already riven by corruption and religious violence.

"It's sort of the nightmare scenario," said Huber. "None of the economists are modeling what happens to a country's GDP if 10 percent of the population is refugees sitting in refugee camps. But look at the real world. What happens if one person who was doing labor in China has to move to Kazakhstan, where they aren't working? In an economic model, they'd be immediately put to work. But in the real world they'd just sit there and get pissed. If people don't have economic hope and they're displaced, they tend to get mad and blow things up. It's the kind of world in which the major institutions, including nations as a whole, have their existence threatened by mass migration. That's where I see things heading by midcentury."

And it doesn't get any better after 2050. But forecasts about the disintegration of society are social and political speculations and have nothing to do with mass extinctions. Huber is more interested in the hard limits of biology. He wants to know when humans themselves will actually start to disintegrate. His 2010 paper on the subject was inspired by a chance meeting with a colleague.

"I presented a paper at a conference about how hot tropical temperatures were in the geological past and [University of New South Wales climate scientist] Steve Sherwood was in the audience. He heard my talk, and he started asking himself the very basic question, 'How hot and humid can it get before things start dying?' It was literally just an order of magnitude kind of question. I guess he thought about it and realized that he didn't know the answer and wasn't sure anyone else did either. . . . Our paper really wasn't motivated by the future climate per se, because when we started we didn't know if there was any kind of realistic future climate state that would fall within this habitability limit. When we started, it was just like, 'We don't know. Maybe you have to go to, like, 50 degrees Celsius global mean temperature.' Then we ran a whole set of model results, and it was rather alarming to us."

Sherwood and Huber calculated their temperature thresholds using the so-called wet-bulb temperature, which basically measures how much you can cool off at a given temperature. If humidity is high, for instance, things like sweat and wind are less effective at cooling you down, and the wet-bulb temperature accounts for this.

"If you take a meteorology class, the wet-bulb temperature is calculated by basically taking a glass thermometer, putting it in a tight wet sock, and swinging it around your head," he said. "So when you assume that this temperature limit applies to a human, you're really kind of imagining a gale force wind, blowing on a naked human being, who's doused in water, and there's no sunlight, and they're immobile, and actually not doing anything other than basal metabolism."

Today the most common maximums for wet-bulb temperatures around the world are 26 to 27 degrees Celsius. Wet-bulb temperatures of 35 degrees Celsius or higher are lethal to humanity. Above this limit, it is impossible for humans to dissipate the heat they generate indefinitely and they die of overheating in a matter of hours, no matter how hard they try to cool off.

"So we were trying to get across the point that physiology and adaptation and these other things will have nothing to do with this limit. It's the E-Z Bake Oven limit," he said. "You cook yourself, very slowly."

What that means is that this limit is likely far too generous for human survivability.

"When you do real modeling, you hit a limit much sooner, because human beings aren't wet socks," he said. According to Huber and Sherwood's modeling, 7 degrees Celsius of warming would begin to render large parts of the globe lethally hot to mammals. Continue warming past that, and truly huge swaths of the planet currently inhabited by humans would exceed 35 degrees Celsius wet-bulb temperatures and would have to be abandoned. Otherwise, the people who live there would be literally cooked to death.

"People are always like, 'Oh, well, can't we adapt?' and you can, to a point," he said. "It's just after that point that I'm talking about."

Already in today's world, heated less than 1 degree Celsius above preindustrial times, heat waves have assumed a new deadly demeanor. In 2003, two hot weeks killed 35,000 people in Europe. It was called a once-in-500-year event. It happened again three years later (497 years ahead of schedule). In 2010, a heat wave killed 15,000 people in Russia. In 2015, nearly 700 people died in Karachi alone from a heat wave that struck Pakistan while many were fasting for Ramadan. But these tragic episodes are barely a shade of what's projected.

"In the near term—2050 or 2070—the Midwest United States is going to be one of the hardest hit," said Huber. "There's a plume of warm, moist air that heads up through the central interior of the US during just the right season, and man, is it hot and sticky. You just add a couple of degrees and it gets *really* hot and sticky. These are thresholds, right? These aren't just like smooth functions. It gets above a certain number and you hurt yourself very badly."

China, Brazil, and Africa face similarly infernal forecasts, while the already sweltering Middle East has what Huber calls "existential problems." The first flickers of this slow-motion catastrophe might be familiar to Europeans struggling to accommodate the tens of thousands of refugees at their borders: the collapse and mass migration of Syrian society came after a punishing four-year drought. Still others have noted that the Hajj, which brings 2 million religious pilgrims to Mecca each year, will be a physically impossible religious obligation to fulfill due to the limits of heat stress in the region in just a few decades.

But for the very worst-case emissions scenarios, heat waves would not merely be a public health crisis, or a "threat multiplier," as the US Pentagon calls global warming. Humanity would have to abandon most of the earth it now inhabits. In their paper, Huber and Sherwood write: "If warmings of 10°C were really to occur in the next three centuries, the area of land likely rendered uninhabitable by heat stress would dwarf that affected by rising sea level."

Huber said, "If you ask any schoolchild, 'What were mammals doing in the age of the dinosaurs?' they'd say they were living underground and coming out at night. Why? Well, heat stress is a very simple explanation. Interestingly, birds have a higher set point temperature—ours is 37 degrees Celsius, birds' is more like 41. So I actually think that's a very deep evolutionary relic right there. Because that wet-bulb temperature was probably maxing out around 41 degrees Celsius in the Cretaceous, not 37."

The uncommonly pleasant climate window of the past 10,000 years has been among the most equable and stable in the past million years. It's within this unusual interval that all of recorded history has occurred. Viewed in time-lapse, the earth would pulsate with glaciers over the past 2.6 million years as it dipped in and out of the ice ages. Then, in the last frame—in the most recent of countless glacial retreats—agriculture, the division of labor, writing, all of ancient history, messianic cults with global appeal, architecture, coastal cities, peer-reviewed science, and the Choco Taco would appear. But this temporary climatic arrangement should be appreciated for what it is—exceedingly fortunate and rare. In a burn-it-all nightmare, Huber's models produce a global wasteland encompassing half the surface area of the planet and almost all the land currently inhabited by humans.

"Based on what we think we know about plants, you'll violate the temperature threshold at which most plants can survive. So you've probably lost most of your plants by that point, and most of your mammals would be either dead or just coming out at night. But, you know, if you're in Siberia, things are pretty good, northern Canada, southern South America, New Zealand—that's where I plan on buying land."

I joked to Huber that when I visited Newfoundland, I should have been busy scouting out real estate. He responded without even an ounce of humor.

"Yeah, that's a good place," he said. "You'll want to be polewards of 45 degrees latitude."

It would be a return to another planet—one far predating the evolutionary history of *Homo sapiens,* when jungles and reptiles ringed the North Pole. But is there really enough fossil fuel in the ground to bring this primeval planet back to life?

"What we say is this is actually a distinct possibility," Huber said. "It's not something that just couldn't possibly happen. Writing the paper, we would have been just as happy to publish it saying, 'This will never happen.' I would sleep better at night if I knew it wasn't going to happen. But we did the math and were like, 'Oh, actually this could totally happen.'"

It would take more than a century of continued profligate fossil fuel use to get anywhere near 7, much less 12 degrees of warming.

But avoiding it will require the goodwill of energy companies to leave 80 percent of their profitable reserves in the ground, and the creation of staggeringly large new sources of carbon-free energy.

In 2015, all the countries of the world met in Paris to negotiate a plan to prevent the planet from warming by 2 degrees by 2100. Despite the rosy assessment of many editorial writers, they failed catastrophically. There are no binding commitments, and countries' adherence to the agreement is voluntary. Though the signatory countries announced their intent to aim for 1.5 degrees of warming, the agreement itself sheepishly acknowledges that if every country met their optimistic emissions pledges, the planet would still easily sail past 2 degrees. But even if they had succeeded at crafting a meaningful 2-degree treaty, it would have meant that the most ambitious plan yet put forward by the world's leaders would limit warming to a level that will wipe out most of the coral reefs and major parts of the rainforest, bring unprecedented heat waves and legions of extinctions, and eventually drown coastal cities around the world. And since the ocean-climate system doesn't pack it in in 2100, warming and sea level rise would persist and indeed increase for hundreds, if not thousands, of years.

As University of Chicago geophysicist David Archer recently commented on the arbitrary goal, "I have a feeling that by the time we get close to 2 degrees Celsius we'll think it's pretty insane we ever thought that was a target to shoot for."

Still, this 2-degree goal is, in fact, extremely ambitious. To reach it—as the world population continues to add billions of souls—fossil fuel use will need to fall to near zero by midcentury at the same time that the world will have to scrape together nearly 30 terawatts of new carbon-free energy, a preposterous amount equivalent to more than double what the world currently consumes, most of which is from fossil fuels. It's why Columbia economist Scott Barrett wrote about the Paris agreement: "The only way the voluntary contributions pledged thus far could achieve the collective 2-degree goal is if a miracle occurs around 2030, some technological breakthrough forcing global emissions to plummet. Even then, the chances of staying within the 2-degree goal are no better than 50-50."

Huber says that, though they're unwilling to admit it publicly, very few climate scientists ("except for a few people in Germany") actually believe that there's any hope of limiting the planet to 2 degrees Celsius warming by the end of the century. But by setting a modest target, we might ensure that, when we miss, the planet warms by only around 4 degrees—rather than, say, being catapulted back to the Eocene. But what kind of prize is hitting 4 degrees? In 2012, the typically buttoned-up World Bank issued a report predicting that a 4-degree warmer world would unleash "heat waves of unprecedented magnitude and duration." It described this world in more detail:

> In this new high-temperature climate regime [in tropical South America, Central Africa, and all the tropical islands of the Pacific], the coolest months are likely to be substantially warmer than the warmest months at the end of the 20th century. In regions such as the Mediterranean, North Africa, the Middle East, and the Tibetan plateau, almost all summer months are likely to be warmer than the most extreme heat waves presently experienced. . . . Stresses on human health, such as heat waves, malnutrition, and decreasing quality of drinking water due to seawater intrusion, have the potential to overburden health-care systems to a point where adaptation is no longer possible.

The most frightening prospect of all comes from what Donald Rumsfeld, in his gnomic way, deemed the "unknown unknowns." When the armies of well-groomed apparatchiks descend on conference rooms to hash out international climate negotiations, they're armed with graphs that plot smooth functions of emissions, temperature rise, and sea level rise—graphs that end at the artificial date of 2100. Increase carbon dioxide by a certain amount, the models say, and temperature and sea level will rise in lockstep in a linear fashion. The fate of the world, then, becomes an easily calculable cost-benefit analysis, one amenable to smug op-eds by economists. The corn belt will shift north by so-and-so degrees latitude, the GDP of certain countries will respond in kind, and it's all very orderly and predictable.

Unfortunately, this is not how the world has tended to behave in the geological past. Throughout the climate swings of the Pleistocene, the ice sheet that covered North America—one larger even than modern Antarctica—did not merely shrink in response to a few degrees of warming. It exploded. Rather than slowly dwindling over thousands of years, these continents of ice sometimes violently disintegrated in spectacles that unfolded over mere centuries. In one rapid collapse 14,000 years ago, called Meltwater Pulse 1A, three Greenlands' worth of ice fell into the sea in icy flotillas, sending sea level soaring 60 feet. The latest International Panel on Climate Change report calls for a half-meter of sea level rise by 2100.

"Sea level in the geologic past was much more responsive to changes in global climate than what IPCC predicts for the year 2100," writes the University of Chicago's David Archer. "Past sea level varied by 10–20 meters (30–60 feet) for each 1C change in the global average temperature. The IPCC business-as-usual forecast for 3C would translate to 20–50 meters (60–150 feet) of sea level rise." The IPCC may well be right about its half-meter prediction by the end of the century.* But it might not be.

* No one doubts that sea levels will eventually rise by several meters due to anthropogenic warming. The only question is about how much it will rise by the IPCC's arbitrary date of 2100.

People don't talk much about what happens after 2100. On the scale of a human lifetime, the affairs of the next century remain hazy and remote fictions. But since the scope of this book is geological, the year 2100 is an insignificant mile marker, and the passage of centuries an indistinguishable blur, unresolvable in the fossil record. For tens of thousands of years beyond 2100, the earth will remain much warmer and totally unlike what it has been for millions of years. Thawing permafrost on land and methane from the deep may eventually add as much carbon to the atmosphere as human contributions, spiking temperatures even further—in a worst-case scenario possibly as high as the Eocene, when reptiles sunbathed in the Arctic Circle.

And the sea will continue to rise as surely as the sun. Summertime temperatures 3 degrees warmer will eventually melt all of Greenland. And if the collapse of the West Antarctic Ice Shelf is as irreversible as ice sheet modelers and the history of past interglacials tell us, then within a couple of centuries much of Florida will drown. So too will Bangladesh, most of the Nile Delta, and New Orleans. In the centuries beyond, if our experiment with the climate goes unchecked, so too will much of New York City, Boston, Amsterdam, Venice, and countless other temporary shelters of humanity—where they'll rest in watery repose for tens, even hundreds, of thousands of years. Civilization has numbered sixty centuries so far, but the next handful may well see the ocean rise more than 200 feet if we burn it all. This isn't that surprising. In the millennia before civilization, the ocean rose up 400 feet from the edge of the continental shelves. Boston was built as a seafaring city, but a few thousand years ago it would have been landlocked more than 200 miles away from the ocean. That the coastline would continue to migrate inland should come as no great surprise. This is what the ocean does in geological time, mocking the putative permanence of our coastal settlements.

But as extreme as all these potential changes to our planet are, do they have anything to do with mass extinctions? As economists and political scientists peer more than a few decades into this wild future that awaits, their forecasts grade into opaque uncertainty. But paleontologists have seen wild times before.

David Jablonski of the University of Chicago is the rare paleontologist who gets to spend his time, not parsing the morphology of early Devonian crinoid anuses or some such arcana, but actually dwelling on the whole history of life—in all its star turns, harrowing tragedies, and macro-evolutionary glory. I came to the University of Chicago to put our species in this context—to get some truly big-picture perspective. I wanted to know what sort of geological legacy we might leave.

Casting directors in search of someone who looks like a scientist could do worse than Jablonski. When I met him at his office, he was sporting tousled hair and a T-shirt adorned with Warholized brachiopods. He talks with an irrepressible energy, unable to get out the next big idea fast enough. But there are sacrifices one must make in dedicating a life to answering the Big Questions. Jablonski has sacrificed tidiness.

"You should avert your eyes when I open my office door," he warned.

I'm ill placed to comment on the disorganized workspaces of others, but Jablonski's office, deep in the Henry Hinds Laboratory at the University of Chicago, has, let's say, a certain Collyer Brothers charm to it. When he opened the door, I saw hundreds of years' worth of academic papers. A narrow path led through the papered canyons to his desk.

"Just pick your way through the rubble here," he said, pushing aside a teetering totem pole of old, yellowing monographs in

French, German, Russian, and Chinese that documented long-forgotten French fieldwork in Gabon in the 1950s, or Russian expeditions to the far-flung corners of the vast Soviet Union. I removed a tome entitled *Les Bivalvia du Danian et du Montien de la Belgique* from an office chair and sat down.

"Sorry about this, I'm in the middle of a really big data push—actually a K-T data push, as it happens," he said about the reams of science literature. It looked like an ironic punishment doled out by an angry librarian god.

Jablonski has a special knack for naming phenomena in his field, like "dead clade walking" and "Lazarus taxa." The latter refers to species that disappear, sometimes for millions of years after a mass extinction, only to return later in earth history. These Lazarus taxa aren't literally resurrected from the dead like their biblical namesake, but instead bide their time in unusual sanctuaries called refugia. These are the rare spots on earth where quirks of the local environment shelter organisms from the wholesale destruction taking place all around. While there's no evidence for an evolutionary bottleneck in the Bronze Age as attested to by the story of Noah, real arks of a sort might have existed throughout earth history in the form of these refugia. The sanctuaries sheltered the shell-shocked and decimated species until they were able to repopulate the world in the ensuing ages. That refugia have never been found in the fossil record may reflect their rarity and geographic tininess.

"So they're kind of like dark matter," Jablonski said. "We think they're there because we can't see them."

I wanted to know, if the sixth major mass extinction of the Phanerozoic is coming, where will the refugia be?

"There won't be many," he said glumly. "The human footprint is truly pervasive, from McMurdo Station to the north coast of Greenland. From submarine habitats to the tops of mountains.

You have metals deposited in remote lakes in the Andes, and of course, in the ocean, plastic is everywhere. So there won't be any places to hide, really. The groups that are going to do the best are the ones that can actually coexist with people as opposed to the ones that can find the last few hidey-holes. But if society collapses, dogs will just go back to being wolves. The genus *canis* will be just fine in the long run.

"But things like ocean acidification are really going to matter," he continued. "That's the key, right? Because of course there's been plenty of warming in the past. But how do clades deal with warming? They move around. But if you've built hotels, and sewage effluents, and you're dynamiting reefs, you can't move around anymore. And of course, if on top of that you then acidify the ocean, you've again removed potential refugia. And so that's the real problem: we're the perfect storm.

"We're not just warming, we're not just pollution, we're not just overexploitation, we're piling it all on simultaneously. That's why it's really inaccurate to argue that because there's been warming in the past that doesn't count now, because it's part of the perfect storm. I think that all mass extinctions work that way. I think it's going to turn out that that's how all the Big Five work—that lots of things go wrong. Say the K-T wouldn't have been as bad if you hadn't had the Deccan Traps erupting, or the Deccan Traps wouldn't have done that much damage if you hadn't dropped a rock out of the sky. But you combine those. The Permian-Triassic is the same way. The Devonian is the same way. The End-Ordovician is the same way. The Triassic-Jurassic—it's these combinatory things. You've got to get away from single-factor explanations. I suspect a lot of the major events in the history of life involve perfect storms. And we're one of them. If we just did one thing, it wouldn't be that big of a deal, but we're doing everything simultaneously as hard and fast as we possibly can."

Despite the trail of devastation in civilization's wake, Jablonski thinks that humans will ultimately prove extremely extinction-resistant.

"There's a couple of reasons for that," he said. "One is that we're very widespread. Another is that you can't beat culture for resisting all kinds of horrific things. What I think is more likely is that quality of life is going to go down the tubes for most humans, not that the species itself is at risk. It would take really focused, concentrated attention to wipe us out. After all, humans did pretty well without industrialized societies for hundreds of thousands of years. But on the other hand, someone like me, who needs glasses, would not have been particularly happy as a Neanderthal. So it's a quality-of-life thing rather than a do-or-die thing for the species, it seems to me."

Jablonski agrees with Erwin that we have not yet even come close to inflicting a major mass extinction on the world.

"No, we're not there for sure," he said. "Right now, statistically speaking, the selectivity of extinction is mostly like background extinction, right? It hinges on things like the geographic ranges of individual species, it hinges on things like trophic level and body size, and other things that are not particularly important selective factors during the Big Five.

"So in that sense, we're still in the background zone, which is good news!" he said, shaking his hands in mock celebration. "The problem is, we're creating this perfect storm and that means that it's not impossible that we'll hit a tipping point."

Back at the diner in New Hampshire, Huber told me about his "favorite story": the US Army's real-life parable of the so-called Motivated Point Man. In 1996, a platoon of light infantry spent days in the Puerto Rican jungle acclimatizing to stifling heat and humidity, cautiously monitoring their water intake before simulating a nighttime raid. The platoon included "some of the most fit and motivated soldiers in the battalion." When the evening of the raid came, the platoon leader began leading his troops through the jungle, machete-ing a path through the brush. Before long, he was felled by fatigue and delegated his leadership to an underling. When the second private failed to advance the platoon quickly enough, the platoon leader demanded to lead again. But soon he found himself hyperthermic and unable to walk. His soldiers had to douse him in cold water and supply him with intravenous infusions. Eventually four soldiers had to carry him. Before long, the extra demands vitiated the entire platoon, all of whom began to fall prey to heat stress. The exercise had to be called off before it became a massacre.

"So I look at that as, if it's nighttime and acclimatized, fit people can just disintegrate into a pool of useless people on stretchers. That's what I see happening to society, to cultures," Huber said. "If you want to know how mass extinctions happen, that's how. So when people talk about the Pleistocene megafauna extinctions and Clovis people, sometimes they act like it's a mystery how these things happen. But it happens in exactly the same way. You have something tearing apart the strongest members, the weaker ones try to fill in the gaps, they're really not strong enough to take it, and the whole thing collapses.

"You want to know how societies collapse?" Huber said. "That's how."

"I'm not too worried about that," said Anders Sandberg about Huber's failing power grid scenario, or the prospect of civilization being toppled by climatic and oceanic chaos. Sandberg is a cheerful Swede whose job is to daydream about the apocalypse and the distant future at Oxford University's Future of Humanity Institute. An unabashed transhumanist, he wears dog tags around his neck that indicate his reservation at a cryogenics lab and matter-of-factly says things like, "Even if I manage to stop my biological aging and eventually upload myself to a computer and have backup copies across the galaxy, sooner or later luck runs out." If it weren't for his affiliation with Oxford, Sandberg's musings, like those of many futurists, can sometimes seem outré to the point of deranged. He would argue that he is only extrapolating trends that have, in a few short decades, made life unrecognizable to the generations of humans that came before.

Where concern over climate change has actually fallen among the American public in the past 15 years (even as the threat grows), researchers working on the more speculative threat of artificial intelligence have found an eager audience for their nightmare visions—especially among Silicon Valley donors. Regarding the heat-stressed power grid, Sandberg said that predictions of its demise follow in the classic tradition of underestimating the power of technological change.

"That's very dependent on the power grid working like it currently does," Sandberg told me. "When I was growing up, I vividly remember one teacher saying, 'Oh, most people in the world have never made a phone call, *and they never will*. That's because there's not enough copper to give phones to everyone in China. By the time I was in high school, of course, fiber optics was coming along. Now I think the majority of humans have cell phones. So the technology change completely blew that prediction out of the water. It was based on a very sensible observation. But copper, it turned out, wasn't the limiting factor. Similarly with power grids, if we get a more stormy world, it's likely that we're going to make more resilient power grids."

His point was well taken given that we were conducting our wide-ranging discussion about the future of the planet over Skype, a technology that would have been difficult even to describe to his telecom-skeptic childhood teacher. Sandberg's expectations for technological change in the next few decades are ambitious even by the typical standards of science fiction. He said that he tries to eat right and exercise in hopes of making it to a biotechnological horizon that will grant him something like immortality. (But, he conceded, "it's still a pretty fair chance that I'm going to die completely normally.") His hopes for technology are nearly limitless—and, to others I would imagine, terrifying—and he wants to live to experience as yet unimagined technologically mediated states of being far more expansive than those allowed by the limited wet meat in our skulls. The human brain was shaped haphazardly by the ruthless (and goal-less) filter of natural selection and the limits of metabolism, but imagine what states of awareness and subjectivity could be achieved with a synthetic brain limited only by the ambition and imagination of its super-intelligent creators. With so much at stake, it's not surprising that Sandberg spends the rest of his time wondering about what could foreclose such an expansive future and destroy the planet. These are the existential risks.

Even if the end is near, Sandberg thinks it will be unlike any of the past major mass extinctions. There exist existential threats with no historical precedent whatsoever, with infinite impact and a complete resistance to estimations of their likelihood. These include such speculative threats as alien invasions, but they also include Sandberg's bugbear, runaway artificial intelligence. He says that, while the "silliness heuristic" prevents many people from taking it seriously, the quickening pace of technological change could bring extinction to our door in the form of silicon rather than carbon dioxide.

"If you focus completely on the climate stuff and ignore superintelligence, we might find that yeah, the Paper Clipper gets us first."

The Paper Clipper?

"My idea with the Paper Clipper is, you have this artificial intelligence and you give it the goal of making paper clips. So it tries to do the actions that maximize the number of paper clips, and it figures out how to make itself smarter, because if it's smarter, it's better at making paper clips. So it makes itself really smart, and it comes up with a foolproof plan to convert Earth into paper clips, and it implements that plan. That's very bad news for us. The problem here is, of course, if I tried to pull the plug on it, it's smart enough that it's figured out a way of stopping me, because if I pull the plug, there will be fewer paper clips in the world and that's bad. So it must overcome any attempt at stopping it or changing its mind. It might even be that the universe has something like Kantian ethics, so a sufficiently smart mind will recognize that it's a moral truth that it shouldn't turn people into paper clips. Unfortunately, if the architecture of AI is such that it just maximizes its utility, and utility is defined by paper clips, it's going to think, 'Hmm, being moral or paper clips? Paper clips!'"

Rather than Frost's fire or ice, perhaps the world will end in paper clips. Or, to paraphrase another great poet on the deadly hubris of a bygone world, "boundless and bare the lone and level paper clips stretch far away."

Obviously, Sandberg's thought experiment is not actually about paper clips, but about any super-intelligent system that can outwit us and whose goals fail to dovetail with human flourishing. His project may seem hopelessly speculative, and I'll admit to falling prey to the so-called silliness heuristic. When compared to the concrete projections of climate change and ocean acidification in the coming decades, I find the thought experiments at best unpersuasive and, at worst, an overhyped distraction from the clear and present danger of climatic and oceanic chaos. But those who are paid to ponder these techno-visions are just as convinced of their menace as conservation biologists and climate modelers are of the coming shocks to the earth system. One thing is clear: with so many dizzying trends accelerating, from possibly beneficial ones like AI, green energy, and biotech to possibly catastrophic ones like warming, acidification, overpopulation, overfishing, spreading dead zones, soil erosion, resource scarcity, deforestation, and bad AI, the next few centuries are utterly unpredictable.

Underlying certain segments of the environmentalist movement is a sort of existential misanthropy, the idea—even hope—that humans will get what they deserve. That getting spat out by Gaia is just recompense for trashing the planet. These sentiments pop up both in unenlightened online comment sections and in the understandably resigned fatalism of working scientists, many of whom will tell you after a few beers, "We're fucked." Indeed, if the worst projections of climate models come to pass, I will admit to feeling a certain fleeting schadenfreude when today's climate change–denying politicians live to see their home states confronted by rising seas and temperatures. That sort of cruel vindication will of course be tempered by the knowledge of the immense misery visited upon their constituents. As Sandberg and others have pointed out, the experiences of conscious creatures like ourselves should really be the only thing worth caring about.

"There's a whole branch of philosophy, axiology, that thinks about what's good, what's valuable," he said. "I think a fairly common view is that there at least needs to be a valuer. You can't have a universe where there's no one home but there's still an amazingly valuable state of affairs. No, for that we need to have minds in the universe that can actually look and see what's good about it. So we better have more minds. If we foul things up, all the stuff that past generations have been striving for will be lost. They were aiming for some indefinite future, and now it doesn't come around and nobody will even remember what they were striving for. And all the good things we could be making will not come around, and neither will the countless lives.

"But the most chilling thing might be that if there is nobody around, there is no value at all. There is suddenly no point to the universe."

If the human project fails in the next few centuries, that failure will foreclose the joys and sorrows of billions of possible lives. It will also waste the sacrifice of legions of dead soldiers, the masterworks of great artists, and the thoughts of great thinkers who committed the ideals of civilization to yellowing pages—pages that will wither away as surely as the leaves. Distant planets will go unexplored and unmarveled. Great symphonies will go unwritten. The stakes are as high as can be imagined.

"This was fun!" said Sandberg as we both fumbled to move our cursors toward the hang-up icon on our computer screens. "Cheers!"

If we do survive the strange death sentence of the Paper Clipper, then humanity will be coping with the decisions we make in the twenty-first century for ages. It might seem silly to care about the well-being of citizens thousands of years from now, but we still commune with the inner lives of people of antiquity. We read their poetry and oratory, marvel at their architecture, and identify with their humanity. As David Archer points out, had the ancient Greeks similarly indulged in a few centuries of reckless environmental engineering, we'd be living not only with their epic poems, ruins, and pottery but with an alien planet of their making as well. Future humans, living in a coastal megalopolis in Baffin Island in the Canadian Arctic, might similarly marvel at the strange ancient culture 5,000 years ago that was fully aware that it was sacrificing the prospects of civilization and the welfare of the living world to satisfy its hunger for burning ancient plants and sea life buried in the rocks. But this far underestimates the eventual legacy of humanity.

At 12,000 years long and counting, our current warm interglacial has already outlasted many past interglacials of the Pleistocene, which careened back into the ice ages after roughly 10,000-year intermissions. Today sunlight in the Northern Hemisphere summers is dimming, approaching levels that in the past have been sufficient to kick off glacial ages that lasted more than 100,000 years. In the next few centuries, this dimming sunlight may well reach a threshold that in recent geological history has summoned glaciers to march across North America, dropping sea level hundreds of feet. That our brief respite has already outlived past interglacials might be a product of human interference in the carbon cycle since the dawn of agriculture. But it might have more to do with the current shape of our orbit, which alternates between being more circular and more elliptical over hundreds of thousands of years. Today our orbit is similar to that of 400,000 years ago, when a more circular orbit enabled a warm interglacial lasting 50,000 years. If the planet merely grazes the threshold for glaciation in the next few thousand years but doesn't dip below it, it might again be 50,000 years until we wobble back into the deep freeze. But this assumes a planet operating without human influence, one that can freeze and thaw as it has done for the past few million years.

Instead, the ice ages will almost certainly not return in the next few thousand years—which is surely a good thing. But the alternative we're creating, a leap into an extreme greenhouse of the sort not seen for tens of millions of years, is not any better.

If humans burn 2,000 gigatons of carbon, as is predicted under business-as-usual scenarios, even the 50,000-year date for glacial inception will dissolve into the heat. Some carbon dioxide will be removed by the ocean on 1,000-year timescales as dead sea life made of calcium carbonate on the ocean floor dissolves like a Tums in the acidifying oceans, allowing the seas to store even more carbon dioxide. But a significant amount will yet remain in the atmosphere. This is the portion of carbon dioxide that will need to be removed by rock weathering, which happens on a timescale of at least 100,000 years. If it's still too hot for the glacial inception in 50,000 years, the next chance to dip back into the freezer would be 130,000 years from now. But if humans burn all their fossil fuels, even this off-ramp to the ice ages will be missed. The world might well have to wait 400,000 years until enough carbon has been drawn down by natural processes to resume the icy trajectory of the Pleistocene. If we manage to hang around that long, perhaps we could hold off this deep freeze by sanely managing our carbon emissions, ramping them up, when needed, just enough to thwart the advance of the glaciers but not so much as to cause a global catastrophe like the one set in motion today. Or maybe our use of fossil fuels will be so profligate, and our foresight so inept, that we'll burn through it all fast enough to invite both crushing warming and rising seas followed by a crazy swing back to the ice ages.

"The idea that humans in a blink of an eye can basically stop the next glacial inception or delay it for the better part of half a million years is incredible to me," climate and ocean modeler Andy Ridgwell told me.

While much of science, and especially geology and astronomy, drive home man's insignificance in the big picture, now we're beginning to talk about a real chunk of geological time. The decisions we make as a civilization in the next several decades might influence the climate twice as far into the future as our species has existed in the past. Nevertheless, no matter what human beings do—even if we hit the gas for the next several centuries and burn every last molecule of coal, oil, and gas we can find—the rocks will weather away, the oceans will turn over, the seafloor will dissolve, the glaciers will advance, the seas will drain, and the world will shiver. Even if we ramp up CO_2 to Eocene levels, push crocodiles and marlin to the Arctic, and send sea levels soaring by more than 200 feet, it will likely all come violently crashing down into an ice age. Whether this ice age returns 130,000 or 400,000 years in the future, the waterlogged ruins of New Orleans, New York, and the Nile Delta will be exposed again, though who knows how much of them, if anything, will still be preserved after these millennia in the deep.

Peter Ward wrote evocatively about this coming ice world after the brief greenhouse in his unsettling book *The Life and Death of Planet Earth*. "From the vantage point of a derelict and forgotten satellite orbiting far out in space, the reflection of our marbled home is as disquieting as it is dazzling: a reflective expanding white," he wrote.

> The glaciers are growing. The sea level that briefly rose at the height of civilization is now dropping, exposing new coastal plains, linking islands, and creating land bridges. Harbors have become meadows. The English Channel and the Bering Strait have become corridors. All the maps have changed. At night the planet no longer glistens with a galaxy of city lights that once stretched from the Arctic to the Southern Ocean. Instead, the Arctic has been abandoned and the Southern Ocean is largely frozen over. The lights that glitter are in a narrower band hugging the equator and midlatitudes. Many are now campfires.

The oceans will require a similarly epic timescale to restore. "Whatever perturbation we eventually engineer, it will take at least 100,000 years for ocean carbonate chemistry to come back to pre-anthropogenic conditions," UC Santa Cruz paleoceanographer James Zachos told me. "That was predicted twenty-five years ago, and with the PETM, we've verified that theory. It does take 100,000 years to restore ocean chemistry."

But the effects on the biosphere will live on far longer. As previous mass extinctions have shown, biology recovers long after ocean chemistry is done sorting itself out. If we do launch from the icehouse of the Pleistocene into a brief Eocene greenhouse, and then back into the ice—like the Ordovician mass extinction in reverse—it may be too much for the biosphere to take. If humans haven't already consummated it first, this might be where the sixth mass extinction truly comes into its own.

"I think the key thing that we learn from these mass extinctions is, the last thing to recover is biology," said Jonathan Payne. "Getting the carbon out of the system takes hundreds of thousands of years. Rebuilding ecosystems takes millions to tens of millions of years. That's actually the stuff that, for that paleontologist who comes back 100 million years from now, will have left the longest mark on the geological record. It will be the extinctions that we cause."

In the tropics of the next interglacial, some half a million years in the future, the water is once again saturated with calcium carbonate. But empty mounds of bacterial stromatolites flourish where coral reefs once hosted technicolor clouds of fish. On land a monotony of rodents, wild dogs, small birds, and weeds own the world. But send Earth around the sun half a million more times and the first contours of the new world begin to take shape. This is the beginning of the next great radiation. This is the rebirth.

Only pure speculation can suggest what creatures will be produced by this next biological explosion. While the remarkable similarity of animals like dolphins and ichthyosaurs implies that evolution tends to repeat itself, the history of life includes some truly bizarre left-hand turns: long-necked reptiles with four independent paddles, giant carnivorous kangaroos, Permian sharks with circular saws in their mouths, flying reptiles the size of small planes. Nothing in the modern ecosystem resembles these animals. It might be possible to sketch the rough outlines of the future biosphere—like one that includes a shallow ocean stocked with new reef builders that precipitate calcium carbonate and, on land, a new suite of predators and prey—but evolution will surely include some surprises. Perhaps wild dogs, millions of years after reverting to wolves in the collapse of civilization, will take advantage of a landscape denuded of large herbivores and grow to the proportions of the gigantic Oligocene beast *Indricotherium* and grasp at tree branches towering overhead. Perhaps pigeons will chaperone us to our demise before growing to become 15-foot-tall flightless foragers. Perhaps seagulls will acquire powerful flesh-tearing beaks and become apex predators, while in the seas, species like cormorants will explode in size and commit more fully to their marine lifestyle, taking on the shape, dimensions, and menacing aspect of mosasaurs. Just as our lineage had to wait 200 million years after the Permian for another chance at the top, perhaps today's feathery descendants of dinosaurs will rule again in the next age. Of course this is wild speculation—the details will be filled in by the endlessly creative nonmind of evolution and happenstance.

"What people always forget is that the top predators after the dinosaurs went under were giant flightless birds, phorusrhacids—basically another branch of dinosaurs," said Jablonski. "Within 10 to 15 million years after the End-Cretaceous, though, you have bats and you have whales and you were getting together a real grazing ecosystem on land, and it really took off. It's amazing. It all took place only 10 to 15 million years later. Now, mind you, when you're talking about humans, 10 million years is unimaginably long. So you can't just say, 'Oh, it will all be okay in the long run,' because we're talking about the *really* long run. It's not a long run that's meaningful to humans."

9

THE LAST EXTINCTION

800 Million Years from Now

What memories for mud to have!

What interesting other kinds of sitting-up mud I met!

—Kurt Vonnegut, 1963

We now hurl ourselves deep into the far future. As in the footstep analogy of the first chapter—where all of human history takes only a few dozen steps to traverse—we've begun to trudge again for hundreds of miles—that is, hundreds of millions of years—forward in time. It's a planet where the climate tantrums of humans, the whiz-bang ingenuity of our machines, and the projects of our civilization are irrelevant. The continents have been rearranged, entire oceans have been consumed and created, and the constellations have been jumbled and scattered across the sky.

Only a handful of spots on earth will count themselves lucky enough to get covered in sediment and then subside into the earth, where they'll remain shielded from erosion and endure the ages unmolested by plate tectonics. These are the fragments of our modern world that have even the slightest chance of leaving any trace in the rocks in the far, far future of geological time. It's heartening to know that, though its prospects for short-term survival are grim, New Orleans—sitting as it does on the business end of a subsiding sedimentary basin—has perhaps the world's greatest potential for long-term preservation. If, in 100 million years, a geologist from somewhere on the tree of life discovers the flattened strata of the Crescent City in the side of a canyon, I can think of worse ways to be represented in the fossil record than by this one data point about humanity. But while the French Quarter's Preservation Hall may well live up to its name, most of our world will not long endure.

Hundreds of millions of years from now, the world map will be as unrecognizable as the creatures that populate its mountain valleys and patrol its reefs. But certain tropes of earth history will reemerge. Pangaea might have been an interesting time in the tectonic history of the planet, but it wasn't unique. It's now thought that many such supercontinents came together and broke apart in the billions of years of deep time, as part of the Wagnerian-sounding "supercontinent cycle." And the older the supercontinent, it turns out, the more it sounds like it was lifted from an old *Transformers* cartoon: there's Rodinia, Nuna, and the supercratons Vaalbara, Superia, and Sclavia. Today the continents are scattered across the globe, but another family reunion has been scheduled possibly as soon as 250 million years in the future.

Although the mid-Atlantic ridge has been pushing us away from Africa and Europe for more than 200 million years, the Atlantic Ocean—much like its ancestor the Iapetus Ocean—is doomed. Right now, in the shadows of deep-sea trenches at the edge of the Caribbean, in Europe off Gibraltar, and in South America off the Falklands, the continents are having their revenge as subduction zones ravenously chew up ocean crust, drawing the sibling continents toward a reunion. These subduction zones are relatively small now, but they will spread, infecting the continental margins.

"Once that happens, it'll start consuming the whole Atlantic Ocean," said Harvard's Francis MacDonald. And once they get going, these subduction zones will be impossible to stop. Like love-struck mutts wolfing down spaghetti in an alleyway, they'll keep chewing up the ocean crust until they reach their lovers on the other side of the ocean.

As it was in the Permian, my beloved New England coastline may yet again become an arid wasteland hundreds of miles from the ocean, marooned in the bleak interior of another supercontinent.

But this is just one model for a future Pangaea. Another prediction, by geophysicist Ross Mitchell and his colleagues at Yale, has the continents still meeting again in a couple hundred million years, but this time over the North Pole. If this happens, who knows what the implications are for complex life, though at first pass it sounds ruinous.

Perhaps, as the continents close in on each other hundreds of millions of years from now, volcanic island chains will once more be shoved up into great mountain ranges to be weathered away, drawing down carbon dioxide and punishing the world with glaciers and cold. Hundreds of millions of years later, when that supercontinent begins to tear apart—filling at first, perhaps, with rift valley lakes inviting strange creatures to their shores—there will be another enormous, earth-killing continental flood basalt on par with the End-Permian, End-Triassic, or End-Cretaceous eruptions.

But long before any of that comes to pass, perhaps an airless continent of rock currently rolling mutely around the solar system will interrupt the procession of life—a mote of dust hitting one sand grain among billions in the void. Whatever form death assumes, the assault will once again all but destroy the new suite of strange creatures—ones shaped by the blind trajectories of evolution and winnowed by an indifferent world over the eons of the far future. But by then, complex life may already be on the ropes.

Though in the very, *very* short term—over the next few centuries—carbon dioxide may dangerously spike from human activity, from a geological perspective the planet is slowly running out of the stuff. The same weathering processes that set the planet's thermostat are ramping up as the sun—in its life cycle as a main sequence star—grows brighter and brighter over its life span.

"Carbon dioxide in the atmosphere achieves a balance between weathering and degassing of CO_2 from the earth," University of Chicago geochemist David Archer told me in a phone conversation. "But if you keep everything else the same and you make the sun brighter, it will boost the hydrological cycle. So more water will wash over the rocks, which will dissolve more rocks, which will carry more carbon down into the earth in the form of calcium carbonate."

Over the life span of the earth, as the sun has brightened, this background weathering rate has grown more intense and carbon dioxide has been steadily dropping from its stifling pre-Cambrian highs to what could be its ice age minima today (again, ignoring humanity's brief injection of CO_2, which will be washed out of the system in short order). Millions of years after today's flash flood of anthropogenic carbon dioxide is taken out of the atmosphere and buried as limestone at the bottom of the ocean, atmospheric carbon dioxide levels will continue dropping. Eventually our ice age will come to an end as the sun grows ever brighter—even as our blanket of carbon dioxide dwindles away. This will lead to a strange world, one that's both hot and without much CO_2. As a result, there won't be much plant life anywhere, or animal life for that matter, since both depend on the stuff. Already, as carbon dioxide has dropped since dinosaur times, plants have evolved new photosynthetic pathways to adapt to this new low-CO_2 regime. These are the so-called C4 plants, plants like grasses and shrubs and cacti. Over the next few hundred million years, these plants will slowly come to dominate this hot, humid, and generally unpleasant world, while many trees and forests, unable to photosynthesize in the carbon-poor atmosphere, will disappear.

The planet will become increasingly shrubby, barren, and brown until about 800 million years from now, when carbon dioxide will drop below 10 parts per million. When this happens, photosynthesis—and thus plant life—will become impossible. When plant life disappears, so too will the animals that depend on them for food and oxygen. The rivers on these barren continents will once more flow to the sea in wide, sloppy, braiding torrents, as they did in the eons before land plants kept them channeled and winding (and as they briefly did after calamities like the Great Dying).

Even if the life-sustaining carbon cycle wasn't petering out, at about the same time it will be getting unbearably hot. As temperatures, even at the poles, top 40 degrees Celsius and hypercanes lash the nearly barren continents, what life remains will burrow and hibernate during the mercilessly hot, months-long Arctic and Antarctic days (to say nothing of the tropics, which will have been long ago forfeited as unspeakable hellscapes). Perhaps some of these polar animals will even grow sails on their backs to dissipate the heat, like *Dimetrodon*. But unlike in the aftermath of the worst mass extinctions, there will be no respite. It will keep getting relentlessly hotter as the sun grows brighter. Plants will continue to disappear, and both CO_2 and oxygen will continue to bleed away. Proteins will unravel and mitochondria will break down, but the winds will grow hotter still. This is the final mass extinction on planet Earth. On some specific day, at some specific hour, the last animal ever will die.

Long after complex life is gone, its memory preserved only in fossils eroding from desolate cliffs, as temperatures top 70 degrees Celsius, even single-celled eukaryotes will die. The late geochemist Siegfried Franck at the Potsdam Institute for Climate Impact Research, in a paper boldly titled "Causes and Timing of Future Biosphere Extinctions," estimated that all this will take place about 1.3 billion years from now. Long after animals have wandered the earth, and even after the dramatic microcosm of the eukaryotic microbe universe is gone, bacteria will inherit the planet for a few hundred million years more, just as it was in the beginning.

We're back in New England now, though it doesn't really matter where we are—there are no coasts on planet Earth. A few supercontinents have come and gone, but with no oceans left to lubricate the plates, plate tectonics have ground to a halt. Volcanoes, where they exist, are apocalyptic flood basalts burbling through the crust. On a red dune overlooking a salt flat more than 300 feet thick and extending thousands of miles, the red sun swells enormously in the sky, though hazily through the Venus-like atmosphere. It's hundreds of degrees out, and the withering toxic haze makes it hard to believe that this was once a verdant world, teeming with biology. The showy pageantry of complex life is long over, and the former splendor of the oceans and jungles is petrified and buried deep underfoot in limestones, in coal tableaus, and in fossils that no one will ever study. In 1.6 billion years, in the face of a pitilessly hostile and temperamental star, conditions on the planet will become so inimical to life, even deep underground, that bacteria will be extinguished. On the other side of this last mass extinction is eternity. Peter Ward and Donald Brownlee have noted the poetic symmetry to this story of life on earth, where multicellular life, eukaryotes, and prokaryotes will take a bow in the reverse order that they first appeared onstage.

And yet, despite this grim forecast, at this particular moment in geological history we couldn't be luckier. Hundreds of millions of years still stretch before us on this planet, and the fact that we've survived all five major mass extinctions—that the earth has managed to support life for billions of years without being destroyed—might be an almost miraculous circumstance. Squandering our good fortune would be not just a civilizational failing but possibly a cosmologically important one.

The history of mass extinctions underscores this luck. In my research, I was struck by a recurring motif in the literature: if, say, Snowball Earth was a little bit more extreme, or the End-Permian volcanism was a little more intense, or the K-T impactor was a little bigger—that is, if all of these events had been only a little worse—we wouldn't be here to talk about them. How is it possible that we survived so many close calls? Perhaps planets shouldn't always be expected to recover from such calamities, and Earth is just exceedingly lucky. Perhaps on other planets there are no survivors around to ask how it was that their planet survived even one of these catastrophes and stayed habitable for billions of years. Perhaps this is why when we've trained our radio telescopes on the stars in search of cosmic friends, we've heard only silence. Perhaps Earth has had an unexplainable, even miraculous, string of good luck. Perhaps, strangest of all, we survived these events only because we're here to ask these questions.

"I can imagine a universe where planets pop like balloons, where they're getting destroyed with a very high probability," Oxford's Anders Sandberg told me. "But it's a big universe, so there are going to be a few random, very lucky planets that haven't popped over millions and billions of years. They're going to be totally unique, they're going to be very weird. But it's a big universe, so they're going to be there. And on some of these planets, observers will evolve and they're going to think, 'Oh, our planet has been around for billions of years, it's a safe universe!' which is, of course, totally wrong, because they have been selected by the fact that their existence depends on their planet being extremely lucky."

"Had Hale-Bopp hit us, there would not be any surface life on this planet," Peter Ward recently told *Nautilus* magazine, referring to the comet that provided a pleasant show in the night sky to earthlings in 1997 but that—at four times the size of the Chicxulub impactor—would have sterilized the planet had its trajectory been slightly different. "We're not just rare, we're lucky."

Perhaps comets like Hale-Bopp hit planets like Earth all the time. The reason, then, that they haven't hit us—and strangely never have—is because on all the planets that do get hit there's no one sitting around afterwards and wondering about it. This is the "observer selection effect," and it has practical applications. If one were to try to estimate, for instance, the likelihood that a Hale-Bopp-sized rock will hit Earth in the near future, a seemingly logical first step would be to look at the geological record to see how often such massive craters appear in Earth's past. Of course this is hopeless because any planet with earth-killing craters in the recent past wouldn't have observers in their wake to notice them. These unseen threats exist in an "anthropic shadow," censored by our very existence. Even if earth-killing asteroids were exceedingly likely and hit planets like ours all the time, the only observers to ask questions about their likelihood would necessarily live on those few planets with an exceptionally good run of luck dodging space rocks. And in a vast and possibly endless universe, plenty of such planets would exist. Estimates of our future survival and of the future habitability of the planet are therefore biased by the fact that we're here to ask the question in the first place. Maybe our survival of all five major mass extinctions says less about the earth's resilience than it does about the bias of our very existence, and of the planet's astronomically good luck. Perhaps by the time you finish this sentence, this nearly impossible run of luck will have ended and we'll have been vaporized by a long-overdue 100-mile-wide asteroid. Or perhaps sometime soon the eruptions of a truly world-ending continental flood basalt will begin to belch. The universe, it turns out, could be vastly more dangerous than we can possibly estimate based on our own, perhaps exceedingly lucky, past.

"Overconfidence becomes very large for very destructive events," Sandberg and his coauthors Nick Bostrom and Milan Ćirković write in an unusually readable paper titled "Anthropic Shadow: Observation Selection Effects and Human Extinction Risks."

> As a consequence, we should have no confidence in historically based probability estimates for events that would certainly extinguish humanity. While this conclusion may seem obvious, it is not widely appreciated. . . . The risks associated with catastrophes such as asteroidal/cometary impacts, supervolcanic episodes, and explosions of supernovae/gamma-ray bursts are based on their observed frequencies. As a result, the frequencies of catastrophes that destroy or are otherwise incompatible with the existence of observers are systematically underestimated.

If this is true—if the earth is an astronomically strange island of life in a hostile universe—we've been given an almost impossibly generous head start. And from whom much has been given, much is expected. No animal in the history of the planet, and possibly in the visible universe, has ever found itself at so consequential a crossroads as ourselves.

I had these intuitions about our cosmic importance validated in a thrilling conversation with UC Santa Cruz cosmologist Anthony Aguirre, whom I sought out to discover what the ultimate theoretical limits to life are in our universe, in the hundreds of trillions of years ahead. Aguirre's long-term cosmological forecast—though austerely constrained by physics—was so mind-bending that it deserves its own book.* The most striking thing he told me, though, was his belief that humanity has a chance to take part, not just in the legacy of our own mortal planet and its provincial solar system, but in the long-term, grandest-scale story of the cosmos. If the geological timescale makes human events seem small, the cosmological timescale is exponentially more humbling: though our planet has only about 800 million years of habitability left, and the life span of our sun only a few billion years beyond that, 100 trillion (or 100,000 billion) years are still left during which life in our galaxy will be possible before the last stars wink out. Aguirre thinks that if humanity escapes the solar system and its impending demise, we'll stretch across the galaxy and into these unknown eons. This is what Arthur C. Clarke had in mind when he wrote, in the best tradition of aspirational science fiction, about the expansive potential of humanity and of the generations of the far future:

> They will know that before them lie, not the millions of years in which we measure the eras of geology, nor the billions of years which span the past lives of the stars, but years to be counted literally in trillions. . . . But for all that, they

* For example, the next mass extinction might strike when the laws of physics go spontaneously berserk.

may envy us, basking in the bright afterglow of Creation;
for we knew the universe when it was young.

But the opportunity to partake in this sweeping *Star Trek*–ian future might depend on our behavior in just the next few decades. Even though Aguirre works with scales and time spans that reinforce the astronomical insignificance of our species, he nevertheless thinks that our stewardship of the planet in the coming years is existentially, even cosmologically, consequential.

"I think we're at the point where essentially—depending on what happens in the next 100 years—I think it's likely that either civilization and potentially all life on earth is going to self-destruct, [or] if it doesn't, I think the likelihood is we will manage to get to nearby planets, then faraway planets, and sort of spread throughout the galaxy," he said. "And so, if you compare those futures, one of them has basically zero interesting conscious stuff going on in it—depending on where you count animals and things—and one of them has an exponentially growing supply of interesting conscious experience. That's a big deal. If we were just one species among many throughout the galaxy, it would kind of be like, 'Well, if we do ourselves in, we had it coming. We got what we deserve.' But if we're kind of the only one in the galaxy—or one of very few—that's a huge future that we've extinguished. And it's all just because we're being stupid now."

I wrote a book about the mortality of the planet at the same time that someone I deeply loved died. My mom. This loss colored for me what was becoming a more and more gloomy evaluation of the prospects for humanity. But my mom never let this gloom in. As she grew ill she marshaled all the consolations of literature and art to her side: the rousing oratory of Henry V's Saint Crispin's

Day speech, the playful vitality of Matisse, the brassy defiance of Elaine Stritch singing Sondheim, the mystic earthiness of Van Morrison—and the comfort of a faith I envied. But when she was near the end, she was fond of quoting the English medieval mystic Julian of Norwich. "All shall be well, and all shall be well and all manner of thing shall be well," she said.

I didn't buy it. The headlines made it daily more difficult to believe. At the same time that the only known habitable planet in the galaxy seems to be careening into geologic catastrophe, beheadings and crucifixions over centuries-old creeds, pandering nativist demagogues, and tribal recriminations dominate the newscasts. As a species, we seem so ill equipped for what lies ahead. Perhaps the trying changes of the coming centuries will prune us of our adolescent stupidity and superstition, and we'll emerge as worthy and capable stewards of the world for millions of years to come. Perhaps we're at the dawn of the Sapiezoic Eon*—a wild flourishing of intelligence and creativity, a new epoch as wondrous and different from the age of animals as that age was from the age of bacteria that preceded it. Or maybe the whole bizarre fever dream of humanity will be nothing more than a single strange sliver of stratigraphy, capped by mass extinction and buried in the canyons of the far future.

When I left Aguirre's office in Santa Cruz, I drove down to the ocean and stood on a ledge made of Pliocene seafloor, which jutted out over the ocean. Underfoot, the rock was pocked with clamshells from millions of years ago. The sky had shed its afternoon pastels, and now an incandescent pink seared the horizon, grading to deep-space dark overhead. Sidewalk lampposts cast the gaudy

* Coined by astrobiologist David Grinspoon.

orange glow of photons from the Cretaceous across the wharf. Similar lights dotted the land still farther down the coast. Hot-blooded cormorants and pelicans—dinosaurs all—crowded close to each other on an island hunk of ancient sandstone that poked up through the waves. All around them, the evening convocation of sea lions was announced by rowdy barking. Closer kin to us, they returned to the sea after the monsters disappeared—to chase the fish that never left. I sat there, at the end of the world, for a long time. The pink sky dissolved, revealing starlight that's raced across the void for eons. The reddening stars tell us that the heavens are flying apart and will someday go dark for an eternity.

In the moonshine, between the silvery tumbling of sea lions, I made out surfers perched on their boards, bobbing up and down on the waves, searching the horizon. The waves came in and went out, as they always have. I don't know why, but I believed my mom: all shall be well.

ACKNOWLEDGMENTS

"If you don't enjoy spending time alone, you won't enjoy writing a book," writes Thomas Ricks. I learned this the hard way. But however isolating the experience can be, (as the acknowledgment section cliché goes) no book is written in isolation. What follows is a very partial list of people who supported me—materially, morally, and otherwise—in bringing this book to life.

Thanks to Hilary Redmon for her initial enthusiasm for the project. To Denise Oswald for her guidance and wisdom in cutting the manuscript down to size and making the final product eminently more readable. Special thanks to Laurie Abkemeier for helping me since the start: for her belief both in the story and in my ability to tell it, as well as for guiding me through the alien landscape of book publishing. To my family, and especially my sister, for reviewing early drafts of chapters. To friends for their support—particularly those, like Sean Mulderrig, with access to paywalled science journals. Thanks to others for putting me up on couches and in spare rooms as I criss-crossed the country. To Dutch Leonard and Julie Wells for encouraging me early in my writing career to find my voice. To the staff at Area Four in Cambridge and Diesel Café in Somerville, Massachusetts, for keeping me well caffeinated.

Special thanks to the many geologists and paleontologists who were so generous with their time, either in informal discussions between sessions at meetings, in e-mails, phone calls, or in trips to rock exposures. I am likely forgetting many people, but a partial list of scientists and interview subjects who helped me include (alphabetically): Anthony Aguirre, Thomas Algeo, David Archer, Richard Bailey, Nina Bednarsek, David Bond, Doug Brezinski, Stephen Brusatte, Simon Darroch, Cole Edwards, Douglas Erwin, Richard Feely, Seth Finnegan, Gener, David Harper, Jonena Hearst, Bill Heimbrock and the members of the Dry Dredgers, Matthew Huber, David Jablonski, Joe Keiper and the Virginia Museum of Natural History, Gerta Keller, Lee Kump, Gary Lash, Stephen Leslie, Cindy Looy, Francis MacDonald, Rowan Martindale, Jay Melosh, Charles Mitchell, Ross Mitchell, Paul Olsen, Jonathan Payne, Mario Rebolledo, Mark Richards, Andy Ridgwell, Doug Rowe, Michael Ryan, Lauren Sallan, Matthew Saltzman, Anders Sandberg, Morgan Schaller, William Stein, Alicia Stigall, Henrik Svensen, Peter Ward, Thomas Williamson, Kristin Wyckoff, James Zachos, Jan Zalasiewicz, and especially Jonathan Knapp, for his generosity of time, energy, and spirit in helping me understand the weird world of the Permian. Special thanks also goes to the signatories of the Treaty of Utrecht for their tireless work in helping to end the War of Spanish Succession. And a special anti-acknowledgment to Microsoft Word for the year of a thousand crashes. Thanks to the developers at Scrivener for inventing a superior product. Finally, none of this would be possible without planet Earth: may your next 600 million years be as lively as the last. Sláinte!

BIBLIOGRAPHY

INTRODUCTION

Bond, David P. G., and Paul B. Wignall. "Large igneous provinces and mass extinctions: An update." *Geological Society of America: Special Papers* 505 (2014).

Dodd, Sarah C., Conall Mac Niocaill, and Adrian R. Muxworthy. "Long duration (> 4 Ma) and steady-state volcanic activity in the early Cretaceous Paraná–Etendeka Large Igneous Province: New palaeomagnetic data from Namibia." *Earth and Planetary Science Letters* 414 (2015): 16–29.

Hazen, Robert M. *The Story of Earth: The First 4.5 Billion Years, from Stardust to Living Planet.* New York: Viking, 2012.

Hönisch, Bärbel, et al. "The geological record of ocean acidification." *Science* 335.6072 (2012): 1058–1063.

Raup, David M. "Biogeographic extinction: A feasibility test." *Geological Society of America: Special Papers* 190 (1982): 277–282.

Taylor, Paul D. *Extinctions in the History of Life.* Cambridge: Cambridge University Press, 2004.

Ward, Peter D. *Under a Green Sky: Global Warming, the Mass Extinctions of the Past, and What They Can Tell Us About Our Future.* New York: Smithsonian/HarperCollins, 2007.

Worm, Boris, et al. "Global patterns of predator diversity in the open oceans." *Science* 309.5739 (2005): 1365–1369.

1. BEGINNINGS

Bailey, R. H., and B. H. Bland. "Ediacaran fossils from the Neoproterozoic Boston Bay Group, Boston area, Massachusetts." *Geological Society of America: Abstracts with Programs* 32 (2000).

Erwin, Douglas H., and Sarah Tweedt. "Ecological drivers of the Ediacaran-Cambrian diversification of Metazoa." *Evolutionary Ecology* 26.2 (2012): 417–433.

Erwin, Douglas H., and James W. Valentine. *The Cambrian Explosion: The Construction of Animal Biodiversity.* New York: W. H. Freeman, 2013.

Laflamme, Marc, et al. "The end of the Ediacara biota: Extinction, biotic replacement, or Cheshire Cat?" *Gondwana Research* 23.2 (2013): 558–573.

Lenton, Timothy M., Richard A. Boyle, Simon W. Poulton, Graham A. Shields-Zhou, and Nicholas J. Butterfield. "Co-evolution of eukaryotes and ocean oxygenation in the Neoproterozoic era." *Nature Geoscience* 7.4 (2014): 257–265. doi:10.1038/ngeo2108.

Williams, Mark, et al. "Is the fossil record of complex animal behaviour a stratigraphical analogue for the Anthropocene?" *Geological Society, London: Special Publications* 395.1 (2014): 143–148.

Zalasiewicz, Jan, et al. "The technofossil record of humans." *Anthropocene Review* 1.1 (2014): 34–43. doi:10.1177/2053019613514953.

2. THE END-ORDOVICIAN MASS EXTINCTION

Armstrong, Howard A., and David A. T. Harper. "An earth system approach to understanding the end-Ordovician (Hirnantian) mass extinction." *Geological Society of America: Special Papers* 505 (2014): 287–300.

Eiler, John M. "Paleoclimate reconstruction using carbonate clumped isotope thermometry." *Quaternary Science Reviews* 30.25 (2011): 3575–3588.

Fortey, Richard. "Olenid trilobites: The oldest known chemoautotrophic symbionts?" *Proceedings of the National Academy of Sciences* 97.12 (2000): 6574–6578.

———. "The lifestyles of the trilobites." *American Scientist* 92 (June 2000): 446–453.

Graham, Alan. *A Natural History of the New World: The Ecology and Evolution of Plants in the Americas.* Chicago: University of Chicago Press, 2011.

Grahn, Yngve, and Stig M. Bergstrom. "Chitinozoans from the Ordovician-Silurian boundary beds in the eastern Cincinnati region in

Ohio and Kentucky." *Ohio Journal of Science* 85.4 (September 1985): 175–183.

Harper, David A. T., Emma U. Hammarlund, and Christian M. Ø. Rasmussen. "End Ordovician extinctions: A coincidence of causes." *Gondwana Research* 25.4 (2014): 1294–1307.

Karabinos, Paul, Heather M. Stoll, and J. Christopher Hepburn. "The Shelburne Falls arc: Lost arc of the Taconic orogeny." In *Guidebook for Field Trips in the Five College Region: 95th Annual Meeting of the New England Intercollegiate Geological Conference, October 10–12, 2003,* edited by John B. Brady and John Thomas Cheney (Northampton, MA: Smith College, Department of Geology, 2003), B3-3–B3-17.

Kröger, Björn. "Cambrian-Ordovician cephalopod palaeogeography and diversity." *Geological Society, London: Memoirs* 38.1 (2013): 429–448.

Kumpulainen, R. A. "The Ordovician glaciation in Eritrea and Ethiopia, NE Africa." *Glacial Sedimentary Processes and Products: International Association of Sedimentologists Special Publication* 39 (2009): 321–342.

Lamsdell, James C., et al. "The oldest described eurypterid: A giant Middle Ordovician (Darriwilian) megalograptid from the Winneshiek Lagerstätte of Iowa." *BMC Evolutionary Biology* 15.1 (2015): 1.

LeHeron, D. P. "The Hirnantian glacial landsystem of the Sahara: A meltwater-dominated system." In *Atlas of Submarine Glacial Landforms: Modern, Quaternary, and Ancient,* edited by J. A. Dowdeswell, M. Canals, M. Jakobsson, B. J. Todd, E. K. Dowdeswell, and K. Hogan, *Geological Society, London: Memoirs* (2016).

Le Heron, Daniel Paul, and James Howard. "Evidence for Late Ordovician glaciation of Al Kufrah Basin, Libya." *Journal of African Earth Sciences* 58.2 (2010): 354–364.

Melchin, Michael J., et al. "Environmental changes in the Late Ordovician–early Silurian: Review and new insights from black shales and nitrogen isotopes." *Geological Society of America Bulletin* 125.11–12 (2013): 1635–1670.

Meyer, David L., and R. A. Davis. *A Sea Without Fish: Life in the Ordovician Sea of the Cincinnati Region.* Bloomington: Indiana University Press, 2009.

Munnecke, Axel, Mikael Calner, David A. T. Harper, and Thomas Servais. "Ordovician and Silurian sea-water chemistry, sea level, and climate: A synopsis." *Palaeogeography, Palaeoclimatology, Palaeoecology* 296.3–4 (2010): 389–413.

Nesvorný, David, et al. "Asteroidal source of L chondrite meteorites." *Icarus* 200.2 (2009): 698–701.

O'Donoghue, James. "The Second Coming." *New Scientist* 198.2660 (2008): 34–37.

Rudkin, David M., et al. "The world's biggest trilobite—Isotelus rex new species from the Upper Ordovician of northern Manitoba, Canada." *Journal of Paleontology* 77.1 (2003): 99–112.

Skehan, James William. *Roadside Geology of Massachusetts.* Missoula, MT: Mountain Press Publishing, 2001.

Upton, John. "Atlantic circulation weakens compared with last thousand years." *Scientific American,* Climate Central, March 24, 2015.

Webby, B. D. *The Great Ordovician Biodiversification Event.* New York: Columbia University Press, 2004.

Young, Seth A., et al. "A major drop in seawater 87Sr/86Sr during the Middle Ordovician (Darriwilian): Links to volcanism and climate?" *Geology* 37.10 (2009): 951–954.

Zalasiewicz, Jan, and Mark Williams. "The Anthropocene: A comparison with the Ordovician-Silurian boundary." *Rendiconti Lincei* 25.1 (2014): 5–12.

3. THE LATE DEVONIAN MASS EXTINCTION

Algeo, Thomas J., et al. "Hydrographic conditions of the Devono-Carboniferous North American Seaway inferred from sedimentary Mo-TOC relationships." *Palaeogeography, Palaeoclimatology, Palaeoecology* 256.3 (2007): 204–230.

Algeo, Thomas J., et al. "Late Devonian oceanic anoxic events and biotic crises: 'Rooted' in the evolution of vascular land plants." *GSA Today* 5.3 (1995): 45.

Alshahrani, Saeed, and James E. Evans. "Shallow-Water Origin of a Devonian Black Shale, Cleveland Shale Member (Ohio Shale), Northeastern Ohio, USA." *Open Journal of Geology* 4.12 (2014): 636.

Botkin-Kowacki, Eva. "Lungs found in mysterious deep-sea fish." *Christian Science Monitor,* September 16, 2015.

Carmichael, Sarah K., et al. "A new model for the Kellwasser Anoxia Events (Late Devonian): Shallow water anoxia in an open oceanic setting in the Central Asian Orogenic Belt." *Palaeogeography, Palaeoclimatology, Palaeoecology* 399 (2014): 394–403.

Clack, Jennifer A. *Gaining Ground: The Origin and Evolution of Tetrapods.* Bloomington: Indiana University Press, 2002.

Dalton, Rex. "The fish that crawled out of the water." *Nature* (April 5, 2006): doi:10.1038/news060403-7.

Friedman, Matt, and Lauren Cole Sallan. "Five hundred million years of extinction and recovery: A Phanerozoic survey of large-scale diversity patterns in fishes." *Palaeontology* 55.4 (2012): 707–742. doi:10.1111/j.1475-4983.2012.01165.x.

Gibling, Martin R., and Neil S. Davies. "Palaeozoic landscapes shaped by plant evolution." *Nature Geoscience* 5.2 (2012): 99–105. doi:10.1038/ngeo1376.

Haddad, Emily Elizabeth. "Paleoecology and geochemistry of the Upper Kellwasser Black Shale and Extinction Event." PhD diss., University of California, Riverside (2015).

McGhee, George R., Jr. *The Late Devonian Mass Extinction: The Frasnian/Famennian Crisis.* New York: Columbia University Press, 1996.

———. *When the Invasion of Land Failed: The Legacy of the Devonian Extinctions.* New York: Columbia University Press, 2013.

———. "The search for sedimentary evidence of glaciation during the Frasnian/Famennian (Late Devonian) biodiversity crisis." *The Sedimentary Record* 12.2 (June 2014): 4–8. http://www.sepm.org/CM_Files/SedimentaryRecord/SedRecord12-2-5.pdf.

Morris, Jennifer L., et al. "Investigating Devonian trees as geo-engineers of past climates: Linking palaeosols to palaeobotany and experimental geobiology." *Palaeontology* 58.5 (2015): 787–801.

Mottequin, Bernard, et al. "Climate change and biodiversity patterns in the Mid-Palaeozoic (Early Devonian to Late Carboniferous)—IGCP 596 (2011–2015)." *Palaeobiodiversity and Palaeoenvironments* 91.2 (2011): 161–162. doi:10.1007/s12549-011-0053-5.

National Science Foundation. "Too much of a good thing: Human activities overload ecosystems with nitrogen." Press release 10-183, October 7, 2010. https://www.nsf.gov/news/news_summ.jsp?cntn_id=117744.

Over, D. Jeffrey. "The Frasnian/Famennian boundary in central and eastern United States." *Palaeogeography, Palaeoclimatology, Palaeoecology* 181.1 (2002): 153–169.

Over, D. J., J. R. Morrow, and P. B. Wignall. *Understanding Late Devonian and Permian-Triassic Biotic and Climatic Events: Towards an Integrated Approach.* Amsterdam: Elsevier, 2005.

Ruddiman, William F., and Ann G. Carmichael. "Pre-industrial depopulation, atmospheric carbon dioxide, and global climate." *Interactions Between Global Change and Human Health (Scripta Varia)* 106 (2006): 158–194.

Scott, Evan E., Matthew E. Clemens, Michael J. Ryan, Gary Jackson, and James T. Boyle. "A Dunkleosteus suborbital from the Cleveland Shale, northeastern Ohio, showing possible Arthrodire-inflicted bite marks: Evidence for agonistic behavior, or postmortem scavenging?" *Geological Society of America: Abstracts with Programs* 44.5 (2012): 61.

Shubin, Neil. *Your Inner Fish: A Journey into the 3.5-Billion-Year History of the Human Body.* New York: Pantheon, 2008.

Stein, William E., Christopher M. Berry, Linda Vanaller Hernick, and Frank Mannolini. "Surprisingly complex community discovered in the Mid-Devonian fossil forest at Gilboa." *Nature* 483.7387 (2012): 78–81. doi:10.1038/nature10819.

Stigall, Alycia L. "Speciation collapse and invasive species dynamics during the Late Devonian 'Mass Extinction.'" *GSA Today* 22.1 (2012): 4–9.

4. THE END-PERMIAN MASS EXTINCTION

Aarnes, Ingrid. "Sill emplacement and contact metamorphism in sedimentary basins." PhD diss., Faculty of Mathematics and Natural Sciences, University of Oslo, 2010.

Algeo, Thomas J., Zhong-Qiang Chen, and David J. Bottjer. "Global review of the Permian-Triassic mass extinction and subsequent recovery: Part II." *Earth-Science Reviews* 149 (2015): 1–4.

Boyer, Diana L., David J. Bottjer, and Mary L. Droser. "Ecological signature of Lower Triassic shell beds of the western United States." *Palaios* 19.4 (2004): 372–380.

Chen, Zhong-Qiang, Thomas J. Algeo, and David J. Bottjer. "Global review of the Permian-Triassic mass extinction and subsequent recovery: Part I." *Earth-Science Reviews* 137 (2014): 1–5.

Clapham, Matthew E. "Extinction: End-Permian Mass Extinction." *eLS* (2013). doi:10.1002/9780470015902.a0001654.pub3.

Cui, Ying, and Lee R. Kump. "Global warming and the end-Permian extinction event: Proxy and modeling perspectives." *Earth-Science Reviews* 149 (2015): 5–22.

Day, Michael O., et al. "When and how did the terrestrial mid-Permian mass extinction occur? Evidence from the tetrapod record of the Karoo Basin, South Africa." *Proceedings of the Royal Society B* 282.1811 (July 8, 2015): doi:10.1098/rspb.2015.0834.

Dutton, A., et al. "Sea-level rise due to polar ice-sheet mass loss during past warm periods." *Science* 349.6244 (2015): aaa4019.

Emanuel, Kerry A., et al. "Hypercanes: A possible link in global extinction scenarios." *Journal of Geophysical Research: Atmospheres* 100.D7 (1995): 13755–13765.

Erwin, Douglas H. *Extinction: How Life on Earth Nearly Ended 250 Million Years Ago.* Princeton, NJ: Princeton University Press, 2006.

Grasby, Stephen E., et al. "Mercury anomalies associated with three extinction events (Capitanian crisis, latest Permian extinction and the Smithian/Spathian extinction) in NW Pangea." *Geological Magazine* 153.2 (2016): 285–297.

Knoll, Andrew H., et al. "Paleophysiology and end-Permian mass extinction." *Earth and Planetary Science Letters* 256.3 (2007): 295–313.

Payne, Jonathan L. "The End-Permian mass extinction and its aftermath: Insights from non-traditional isotope system." Geological Society of America annual meeting, Vancouver, British Columbia (2014).

Payne, Jonathan L., and Matthew E. Clapham. "End-Permian mass extinction in the oceans: An ancient analog for the twenty-first century?" *Annual Review of Earth and Planetary Sciences* 40 (2012): 89–111.

Peltzer, Edward T., and Peter G. Brewer. "Beyond pH and temperature: Thermodynamic constraints imposed by global warming and ocean acidification on mid-water respiration by marine animals." Theme Session Question 6. International Council for the Exploration of the Sea (ICES) Annual Science Conference, September 22–26, 2008, Halifax, Nova Scotia.

Retallack, Gregory J. "Permian and Triassic greenhouse crises." *Gondwana Research* 24.1 (2013): 90–103.

Retallack, Gregory J., Roger M. H. Smith, and Peter D. Ward. "Vertebrate extinction across Permian-Triassic boundary in Karoo Basin, South Africa." *Geological Society of America Bulletin* 115.9 (2003): 1133–1152.

Rey, Kévin, et al. "Global climate perturbations during the Permo-Triassic mass extinctions recorded by continental tetrapods from South Africa." *Gondwana Research* 37 (September 2015): 384–396.

Schneebeli-Hermann, Elke, et al. "Evidence for atmospheric carbon injection during the end-Permian extinction." *Geology* 41.5 (2013): 579–582.

Schubert, Jennifer K., and David J. Bottjer. "Aftermath of the Permian-Triassic mass extinction event: Paleoecology of Lower Triassic carbonates in the western USA." *Palaeogeography, Palaeoclimatology, Palaeoecology* 116.1 (1995): 1–39.

Sephton, M. A., H. Visscher, C. V. Looy, A. B. Verchovsky, and J. S. Watson. "Chemical constitution of a Permian-Triassic disaster species." *Geology* 37.10 (2009): 875–878. doi:10.1130/G30096A.1.

Smith, Roger M. H., and Peter D. Ward. "Pattern of vertebrate extinctions across an event bed at the Permian-Triassic boundary in the Karoo Basin of South Africa." *Geology* 29.12 (2001): 1147–1150.

Svensen, Henrik, Alexander G. Polozov, and Sverre Planke. "Sill-induced evaporite- and coal-metamorphism in the Tunguska Basin, Siberia, and the implications for end-Permian environmental crisis." *European Geosciences Union General Assembly Conference Abstracts* 16 (2014).

Svensen, Henrik, et al. "Siberian gas venting and the end-Permian environmental crisis." *Earth and Planetary Science Letters* 277.3 (2009): 490–500.

Tabor, Neil J. "Wastelands of tropical Pangea: High heat in the Permian." *Geology* 41.5 (2013): 623–624.

Ward, Peter D., David R. Montgomery, and Roger Smith. "Altered river morphology in South Africa related to the Permian-Triassic extinction." *Science* 289.5485 (2000): 1740–1743.

Ward, Peter D., et al. "Abrupt and gradual extinction among Late Permian land vertebrates in the Karoo Basin, South Africa." *Science* 307.5710 (2005): 709–714.

Wignall, Paul B. "Volcanism and mass extinctions." *Volcanoes and the Environment* (2005): 207–226.

5. THE END-TRIASSIC MASS EXTINCTION

Blackburn, Terrence J., et al. "Zircon U-Pb geochronology links the end-Triassic extinction with the Central Atlantic Magmatic Province." *Science* 340.6135 (2013): 941–945.

Cuffey, Roger J., et al. "Geology of the Gettysburg battlefield: How Mesozoic events and processes impacted American history." *Field Guides* 8 (2006): 1–16.

Fernand, Liam, and Peter Brewer, eds. "Report of the workshop on the significance of changes in surface CO_2 and ocean pH in ICES shelf sea ecosystems." International Council for the Exploration of the Sea, London, May 2–4, 2007.

Fraser, Nicholas C. *Dawn of the Dinosaurs: Life in the Triassic.* Bloomington: Indiana University Press, 2006.

Knell, Simon J. *The Great Fossil Enigma: The Search for the Conodont Animal.* Bloomington: Indiana University Press, 2012.

Lau, Kimberly V., et al. "Marine anoxia and delayed Earth system recovery after the end-Permian extinction." *Proceedings of the National Academy of Sciences* 113.9 (2016): 2360–2365.

McElwain, J. C. "Fossil plants and global warming at the Triassic-Jurassic boundary." *Science* 285.5432 (1999): 1386–1390. doi:10.1126/science.285.5432.1386.

Mussard, Mickaël, et al. "Modeling the carbon-sulfate interplays in climate changes related to the emplacement of continental flood basalts." *Geological Society of America: Special Papers* 505 (2014): 339–352.

Olsen, Paul E. "Paleontology and paleoecology of the Newark Supergroup (early Mesozoic, eastern North America)." In *Triassic-Jurassic Rifting: Continental Breakup and the Origins of the Atlantic Ocean and Passive Margins,* edited by W. Manspeizer (Amsterdam: Elsevier, 1988), 185–230.

Olsen, Paul E., and Emma C. Rainforth. "'The Age of Dinosaurs' in the Newark Basin, with special reference to the Lower Hudson Valley."

In *New York State Geological Association Guidebook* (New York State Geological Association, 2001), 59–176.

Olsen, Paul E., Jessica H. Whiteside, and Philip Huber. "Causes and consequences of the Triassic–Jurassic mass extinction as seen from the Hartford basin." In *Guidebook for Field Trips in the Five College Region: 95th Annual Meeting of the New England Intercollegiate Geological Conference, October 10–12, 2003,* edited by John B. Brady and John Thomas Cheney (Northampton, MA: Smith College, Department of Geology, 2003), B5-1–B5-41.

Pálfy, József, and Ádám T. Kocsis. "Volcanism of the Central Atlantic magmatic province as the trigger of environmental and biotic changes around the Triassic-Jurassic boundary." *Geological Society of America: Special Papers* 505 (2014): 245–261.

Pieńkowski, Grzegorz, Grzegorz Niedźwiedzki, and Paweł Brański. "Climatic reversals related to the Central Atlantic magmatic province caused the end-Triassic biotic crisis: Evidence from continental strata in Poland." *Geological Society of America: Special Papers* 505 (2014): 263–286.

Schaller, Morgan F., James D. Wright, and Dennis V. Kent. "Atmospheric pCO_2 perturbations associated with the Central Atlantic magmatic province." *Science* 331.6023 (2011): 1404–1409.

Steinthorsdottir, Margret, Andrew J. Jeram, and Jennifer C. McElwain. "Extremely elevated CO_2 concentrations at the Triassic/Jurassic boundary." *Palaeogeography, Palaeoclimatology, Palaeoecology* 308.3–4 (2011): 418–432.

Sun, Yadong, Paul B. Wignall, Michael M. Joachimski, David P. G. Bond, Stephen E. Grasby, Xulong Lina Lai, L. N. Wang, Zetian T. Zhang, and Si Sun. "Climate warming, euxinia, and carbon isotope perturbations during the Carnian (Triassic) Crisis in South China." *Earth and Planetary Science Letters* 444 (June 15, 2016): 88–100.

Sun, Yadong, et al. "Lethally hot temperatures during the Early Triassic greenhouse." *Science* 338.6105 (2012): 366–370.

Veron, J. E. N. *A Reef in Time: The Great Barrier Reef from Beginning to End.* Cambridge, MA: Belknap Press of Harvard University Press, 2008.

Whiteside, Jessica H., et al. "Insights into the mechanisms of end-Triassic mass extinction and environmental change: An integrated paleontologic, biomarker, and isotopic approach." Geological Society of America annual meeting, Vancouver, British Columbia (2014).

Wignall, P. B. *The Worst of Times: How Life on Earth Survived Eighty Million Years of Extinctions.* Princeton, NJ: Princeton University Press, 2016.

Zanno, Lindsay E., Susan Drymala, Sterling J. Nesbitt, and Vincent P. Schneider. "Early crocodylomorph increases top tier predator diversity during rise of dinosaurs." *Scientific Reports* 5 (2015): 9276. doi:10.1038/srep09276.

6. THE END-CRETACEOUS MASS EXTINCTION

Alvarez, Luis, Walter Alvarez, Frank Asaro, and Helen V. Michel. "Extraterrestrial cause for the Cretaceous-Tertiary extinction." *Science* 208.4448 (1980): 1095–1108.

Alvarez, Walter. *T. Rex and the Crater of Doom.* Princeton, NJ: Princeton University Press, 2013.

Archibald, J. David. "What the dinosaur record says about extinction scenarios." *Geological Society of America: Special Papers* 505 (2014): 213–224.

Belcher, Claire M., et al. "An experimental assessment of the ignition of forest fuels by the thermal pulse generated by the Cretaceous-Palaeogene impact at Chicxulub." *Journal of the Geological Society* 172.2 (2015): 175–185.

Belcher, Claire M., et al. "Geochemical evidence for combustion of hydrocarbons during the KT impact event." *Proceedings of the National Academy of Sciences* 106.11 (2009): 4112–4117.

Bhatia, Aatish. "The Sound So Loud That It Circled the Earth Four Times." Nautilus, September 29, 2014. http://nautil.us/blog/the-sound-so-loud-that-it-circled-the-earth-four-times.

Blonder, Benjamin, et al. "Plant ecological strategies shift across the Cretaceous-Paleogene boundary." *PLoS Biology* 12.9 (2014): e1001949.

Browne, Malcolm W. "The debate over dinosaur extinctions takes an unusually rancorous turn." *New York Times,* January 18, 1988.

Brusatte, Stephen L., Richard J. Butler, Paul M. Barrett, Matthew T. Carrano, David C. Evans, Graeme T. Lloyd, Philip D. Mannion, Mark A. Norell, Daniel J. Peppe, Paul Upchurch, and Thomas E. Williamson. "The extinction of the dinosaurs." *Biological Reviews* 90.2 (2014): 628–642.

Bryant, Edward. *Tsunami: The Underrated Hazard.* New York: Cambridge University Press, 2001.

Chenet, Anne-Lise, et al. "Determination of rapid Deccan eruptions across the Cretaceous-Tertiary boundary using paleomagnetic secular variation: Results from a 1,200-m-thick section in the Mahabaleshwar escarpment." *Journal of Geophysical Research: Solid Earth* 113.B4 (2008).

Coccioni, Rodolfo, Simonetta Monechi, and Michael R. Rampino. "Cretaceous-Paleogene boundary events." *Palaeogeography, Palaeoclimatology, Palaeoecology* 255.1 (2007): 1–3.

Courtillot, Vincent, and Frédéric Fluteau. "A review of the embedded time scales of flood basalt volcanism with special emphasis on dramatically short magmatic pulses." *Geological Society of America: Special Papers* 505 (2014): SPE505–SPE515.

Darwin, Charles. *Works of Charles Darwin: Journal of Researches into the Natural History and Geology of the Countries Visited During the Voyage of HMS* Beagle *Round the World.* Vol. 1. London: John Murray, 1860.

Elbra, T. "The Chicxulub impact structure: What does the Yaxcopoil-1 drill core reveal?" American Geophysical Union Meeting of Americas, Cancún, Mexico, May 2013.

Glen, William. *The Mass-Extinction Debates: How Science Works in a Crisis.* Stanford, CA: Stanford University Press, 1994.

Gulick, Sean. "The 65.5 million year old Chicxulub impact crater: Insights into planetary processes, extinction, and evolution" (lecture). The Austin Forum, Austin, TX, 2013.

Jagoutz, Oliver, et al. "Anomalously fast convergence of India and Eurasia caused by double subduction." *Nature Geoscience* 8.6 (2015): 475–478.

Keller, Gerta. "The Cretaceous-Tertiary mass extinction: Theories and controversies." *Society for Sedimentary Geology (SEPM): Special Publications* 100 (2011): 7–22.

———. "Deccan volcanism, the Chicxulub impact, and the End-Cretaceous mass extinction: Coincidence? Cause and effect?" *Geological Society of America: Special Papers* (2014): 57–89.

Kennett, Douglas J., et al. "Development and disintegration of Maya political systems in response to climate change." *Science* 338.6108 (2012): 788–791.

Kort, Eric A., et al. "Four Corners: The largest US methane anomaly viewed from space." *Geophysical Research Letters* 41.19 (2014): 6898–6903.

Lüders, Volker, and Karen Rickers. "Fluid inclusion evidence for impact-related hydrothermal fluid and hydrocarbon migration in Cretaceous sediments of the ICDP-Chicxulub drill core Yax-1." *Meteoritics and Planetary Science* 39.7 (2004): 1187–1197.

Manga, Michael, and Emily Brodsky. "Seismic triggering of eruptions in the far field: Volcanoes and geysers." *Annual Review of Earth and Planetary Sciences* 34 (2006): 263–291.

Masson, Marilyn A. "Maya collapse cycles." *Proceedings of the National Academy of Sciences* 109.45 (2012): 18237–18238.

———. *Kukulcan's Realm: Urban Life at Ancient Mayapán.* Boulder: University of Colorado Press, 2014.

Napier, W. M. "The role of giant comets in mass extinctions." *Geological Society of America: Special Papers* 505 (2014): 383–395.

Norris, R. D., A. Klaus, and D. Kroon. "Mid-Eocene deep water, the Late Palaeocene thermal maximum, and continental slope mass wasting during the Cretaceous-Palaeogene impact." *Geological Society, London: Special Publications* 183.1 (2001): 23–48.

Oldroyd, D. R. *The Earth Inside and Out: Some Major Contributions to Geology in the Twentieth Century.* London: Geological Society, 2002.

Prasad, Guntupalli V. R., and Ashok Sahni. "Vertebrate fauna from the Deccan volcanic province: Response to volcanic activity." *Geological Society of America: Special Papers* 505 (2014): SPE505–SPE509.

Punekar, Jahnavi, Paula Mateo, and Gerta Keller. "Effects of Deccan volcanism on paleoenvironment and planktic foraminifera: A global survey." *Geological Society of America: Special Papers* 505 (2014): 91–116.

Renne, Paul R., et al. "Time scales of critical events around the Cretaceous-Paleogene boundary." *Science* 339.6120 (2013): 684–687.

Richards, Mark A., et al. "Triggering of the largest Deccan eruptions by the Chicxulub impact." *Geological Society of America Bulletin* 127.11–12 (2015): 1507–1520.

Robinson, Nicole, et al. "A high-resolution marine 187 Os/188 Os record for the late Maastrichtian: Distinguishing the chemical fingerprints of Deccan volcanism and the KP impact event." *Earth and Planetary Science Letters* 281.3 (2009): 159–168.

Samant, Bandana, and Dhananjay M. Mohabey. "Deccan volcanic eruptions and their impact on flora: Palynological evidence." *Geological Society of America: Special Papers* 505 (2014): SPE505–SPE508.

Schoene, Blair, et al. "U-Pb geochronology of the Deccan Traps and relation to the end-Cretaceous mass extinction." *Science* 347.6218 (2015): 182–184.

Smit, J., et al. "Stratigraphy and sedimentology of KT clastic beds in the Moscow Landing (Alabama) outcrop: Evidence for impact related earthquakes and tsunamis." In *New Developments Regarding the KT Event and Other Catastrophes in Earth History.* LPI Contribution 825. Houston: Lunar and Planetary Institute, 1994.

Spicer, Robert A., and Margaret E. Collinson. "Plants and floral change at the Cretaceous-Paleogene boundary: Three decades on." *Geological Society of America: Special Papers* 505 (2014): SPE505.

Swisher, Kevin. "Cretaceous crash." *Texas Monthly* (September 1992): 96–100.

Turner, Billie L., and Jeremy A. Sabloff. "Classic Period collapse of the Central Maya Lowlands: Insights about human-environment relationships for sustainability." *Proceedings of the National Academy of Sciences* 109.35 (2012): 13908–13914.

Wilkinson, David M., Euan G. Nisbet, and Graeme D. Ruxton. "Could methane produced by sauropod dinosaurs have helped drive Mesozoic climate warmth?" *Current Biology* 22.9 (2012): R292–R293.

Wilson, Gregory P. "Mammalian faunal dynamics during the last 1.8 million years of the Cretaceous in Garfield County, Montana." *Journal of Mammalian Evolution* 12.1–2 (2005): 53–76.

———. "Mammals across the K/Pg boundary in northeastern Montana, USA: Dental morphology and body-size patterns reveal extinction selectivity and immigrant-fueled ecospace filling." *Paleobiology* 39.03 (2013): 429–469.

———. "Mammalian extinction, survival, and recovery dynamics across the Cretaceous-Paleogene boundary in northeastern Montana, USA." *Geological Society of America: Special Papers* 503 (2014): 365–392.

Wilson, Gregory P., David G. DeMar, and Grace Carter. "Extinction and survival of salamander and salamander-like amphibians across the Cretaceous-Paleogene boundary in northeastern Montana, USA." *Geological Society of America: Special Papers* 503 (2014): 271–297.

Zongker, Doug. "Chicken Chicken Chicken: Chicken Chicken." *Annals of Improbable Research* (2006). https://isotropic.org/papers/chicken.pdf.

Zürcher, Lukas, and David A. Kring. "Hydrothermal alteration in the core of the Yaxcopoil-1 borehole, Chicxulub impact structure, Mexico." *Meteoritics and Planetary Science Archives* 39.7 (2004): 1199–1221.

7. THE END-PLEISTOCENE MASS EXTINCTION

Brahic, Catherine. "Travel back in time to an Arctic heatwave." *New Scientist,* July 15, 2015.

Hallam, A. *Catastrophes and Lesser Calamities: The Causes of Mass Extinctions.* Oxford: Oxford University Press, 2004.

Harrabin, Roger. "World wildlife populations halved in 40 years." *BBC News,* September 30, 2014.

Hönisch, Bärbel, et al. "Atmospheric carbon dioxide concentration across the mid-Pleistocene transition." *Science* 324.5934 (2009): 1551–1554.

Kent, Dennis V., and Giovanni Muttoni. "Equatorial convergence of India and early Cenozoic climate trends." *Proceedings of the National Academy of Sciences* 105.42 (2008): 16065–16070.

Koch, Paul L. "Land of the lost." *Science* 311.5763 (2006): 957.

Koch, Paul L., and Anthony D. Barnosky. "Late Quaternary extinctions: State of the debate." *Annual Review of Ecology, Evolution, and Systematics* (2006): 215–250.

Lenton, Tim, and A. J. Watson. *Revolutions That Made the Earth.* Oxford: Oxford University Press, 2011.

Martin, Paul S. *Twilight of the Mammoths: Ice Age Extinctions and the Rewilding of America.* Berkeley: University of California Press, 2005.

Owen, James. "Farming claims almost half earth's land, new maps show." *National Geographic,* December 9, 2005.

Pearce, Fred. "Global extinction rates: Why do estimates vary so wildly?" *Yale Environment 360* (Yale School of Forestry and Environmental Studies), August 17, 2015. http://e360.yale.edu/feature/global_extinction_rates_why_do_estimates_vary_so_wildly/2904/.

Prothero, Donald R. *Greenhouse of the Dinosaurs: Evolution, Extinction, and the Future of Our Planet.* New York: Columbia University Press, 2009.

Schlosser, C. Adam, Kenneth Strzepek, Xiang Gao, Charles Fant, Élodie Blanc, Sergey Paltsev, Henry Jacoby, John Reilly, and Arthur Gueneau. "The future of global water stress: An integrated assessment." *Earth's Future* 2.8 (2014): 341–61.

Secord, Ross, et al. "Evolution of the earliest horses driven by climate change in the Paleocene-Eocene thermal maximum." *Science* 335.6071 (2012): 959–962.

Stone, Richard. *Mammoth: The Resurrection of an Ice Age Giant.* Cambridge, MA: Perseus Publishing, 2001.

8. THE NEAR FUTURE

Archer, David. *The Long Thaw: How Humans Are Changing the Next 100,000 Years of Earth's Climate.* Princeton, NJ: Princeton University Press, 2009.

Barnosky, Anthony D., et al. "Has the Earth's sixth mass extinction already arrived?" *Nature* 471.7336 (2011): 51–57.

Bostrom, Nick. "Existential risks." *Journal of Evolution and Technology* 9.1 (2002): 1–31.

Brook, Barry W., Navjot S. Sodhi, and Corey J. A. Bradshaw. "Synergies among extinction drivers under global change." *Trends in Ecology and Evolution* 23.8 (2008): 453–460.

Ćirković, Milan M., Anders Sandberg, and Nick Bostrom. "Anthropic shadow: Observation selection effects and human extinction risks." *Risk Analysis* 30.10 (2010): 1495–1506.

Davis, Steven J., et al. "Rethinking wedges." *Environmental Research Letters* 8.1 (2013): 011001.

DeConto, Robert M., and David Pollard. "Contribution of Antarctica to past and future sea-level rise." *Nature* 531.7596 (2016): 591–597.

Dirzo, Rodolfo, et al. "Defaunation in the Anthropocene." *Science* 345.6195 (2014): 401–406.

Hansen, James, et al. "Ice melt, sea level rise, and superstorms: Evidence from paleoclimate data, climate modeling, and modern observations that 2 C global warming is highly dangerous." *Atmospheric Chemistry and Physics: Discussion Papers* 15 (2015): 20059–20179.

Hoffert, Martin I. "Farewell to fossil fuels?" *Science* 329.5997 (2010): 1292–1294.

Jagniecki, Elliot A., et al. "Eocene atmospheric CO_2 from the nahcolite proxy." *Geology* 43.12 (2015): 1075–1078.

Lewis, Nathan S. "Powering the planet." *MRS Bulletin* 32.10 (2007): 808–820.

Mann, Michael E., and Lee R. Kump. *Dire Predictions: Understanding Climate Change.* 2nd ed. London: DK, 2015.

Matthews, H. Damon, and Ken Caldeira. "Stabilizing climate requires near-zero emissions." *Geophysical Research Letters* 35.4 (2008).

McInerney, Francesca A., and Scott L. Wing. "The Paleocene-Eocene thermal maximum: A perturbation of carbon cycle, climate, and biosphere with implications for the future." *Annual Review of Earth and Planetary Sciences* 39 (2011): 489–516.

Muhs, Daniel R., et al. "Quaternary sea-level history of the United States." *Developments in Quaternary Sciences* 1 (2003): 147–183.

Muhs, Daniel R., et al. "Sea-level history of the past two interglacial periods: New evidence from U-series dating of reef corals from south Florida." *Quaternary Science Reviews* 30.5 (2011): 570–590.

Pamlin, Dennis, and Stuart Armstrong. "Global challenges: 12 risks that threaten human civilisation—The case for a new category of risks." *Global Challenges Foundation* (February 2015). http://globalchallenges.org/wp-content/uploads/12-Risks-with-infinite-impact-full-report-1.pdf.

Sherwood, Steven C., and Matthew Huber. "An adaptability limit to climate change due to heat stress." *Proceedings of the National Academy of Sciences* 107.21 (2010): 9552–9555.

Sonna, Larry A. "Practical medical aspects of military operations in the heat." *Medical Aspects of Harsh Environments* 1 (2001).

Tollefson, Jeff. "The 8,000-year-old climate puzzle." *Nature* (March 25, 2011): doi:10.1038/news.2011.184.

Zeebe, Richard E., and James C. Zachos. "Long-term legacy of massive carbon input to the Earth system: Anthropocene versus Eocene." *Philosophical Transactions of the Royal Society of London A: Mathematical, Physical, and Engineering Sciences* 371.2001 (2013): 20120006.

Zeliadt, Nicholette. "Profile of David Jablonski." *Proceedings of the National Academy of Sciences* 110.26 (2013): 10467–10469.

9. THE LAST EXTINCTION

Bennett, S. Christopher. "Aerodynamics and thermoregulatory function of the dorsal sail of Edaphosaurus." *Paleobiology* 22.04 (1996): 496–506.

Berner, Robert A., and Zavareth Kothavala. "GEOCARB III: A revised model of atmospheric CO_2 over Phanerozoic time." *American Journal of Science* 301.2 (2001): 182–204.

Evans, D. A. D. "Reconstructing pre-Pangean supercontinents." *Geological Society of America Bulletin* 125.11–12 (2013): 1735–1751.

Franck, S., C. Bounama, and W. Von Bloh. "Causes and timing of future biosphere extinctions." *Biogeosciences* 3.1 (2006): 85–92.

Royer, Dana L., et al. "CO_2 as a primary driver of Phanerozoic climate." *GSA Today* 14.3 (2004): 4–10.

Smith, Kerri. "Supercontinent Amasia to take North Pole position." *Nature* (February 8, 2012). http://www.nature.com/news/supercontinent-amasia-to-take-north-pole-position-1.9996.

Ward, Peter D., and Donald Brownlee. *The Life and Death of Planet Earth: How the New Science of Astrobiology Charts the Ultimate Fate of Our World.* New York: Henry Holt & Co./Times Books, 2003.

Zalasiewicz, J. A., and Kim Freedman. *The Earth After Us: What Legacy Will Humans Leave in the Rocks?* Oxford: Oxford University Press, 2008.

INDEX

DATE			